# Computer Communications and Networks

For other titles published in this series, go to
www.springer.com/series/4198

The **Computer Communications and Networks** series is a range of textbooks, monographs and handbooks. It sets out to provide students, researchers and non-specialists alike with a sure grounding in current knowledge, together with comprehensible access to the latest developments in computer communications and networking.

Emphasis is placed on clear and explanatory styles that support a tutorial approach, so that even the most complex of topics is presented in a lucid and intelligible manner.

Massimo Cafaro · Giovanni Aloisio

Editors

# Grids, Clouds and Virtualization

 Springer

*Editors*
Dr. Massimo Cafaro
Dipartimento di Ingegneria
dell'Innovazione
Università del Salento
Via per Monteroni
73100 Lecce
Italy
massimo.cafaro@unisalento.it

Prof. Giovanni Aloisio
Dipartimento di Ingegneria
dell'Innovazione
Università del Salento
Via per Monteroni
73100 Lecce
Italy
giovanni.aloisio@unisalento.it

*Series Editor*
Professor A.J. Sammes, BSc, MPhil, PhD,
FBCS, CEng
Centre for Forensic Computing
Cranfield University
DCMT, Shrivenham
Swindon SN6 8LA
UK

ISSN 1617-7975
ISBN 978-0-85729-048-9                    e-ISBN 978-0-85729-049-6
DOI 10.1007/978-0-85729-049-6
Springer London Dordrecht Heidelberg New York

British Library Cataloguing in Publication Data
A catalogue record for this book is available from the British Library

Library of Congress Control Number: 2010936502

*Cover design*: VTEX, Vilnius

Printed on acid-free paper

Springer is part of Springer Science+Business Media (www.springer.com)

# Preface

For more than a decade, the development of grid computing was driven by scientific applications. The need to solve large-scale, increasingly complex problems motivated research on grids systems. Many interesting problems have been solved with the help of grids, for instance, the nug30 Quadratic Assignment Problem.

This challenging optimization problem was posed in 1968 and requires, given a set of $n$ facilities, a set of $n$ locations, a distance specified for each pair of locations, and a flow (weight) specified for each pair of facilities (e.g., the amount of supplies transported between the two facilities), assigning all 30 facilities to the 30 different locations with the goal of minimizing the sum of the distances multiplied by the corresponding flows.

Despite its apparent simplicity, the problem is NP-Hard, and the number of possible assignments is extremely large, so that even if you could check a trillion assignments per second, this process would take over 100 times the age of the universe. However, once the algorithms and software necessary to tackle the previously unsolved problem on a computational grid were developed, solving the problem required nearly a week, with a computational endeavor involving more than 1,000 computational resources working simultaneously at eight institutions geographically distributed in different parts of the world.

The FightAIDS@Home project, which is based on the volunteered computing power of the World Community Grid, aims at testing candidate compounds against the variations (or "mutants") of HIV that can arise and cause drug resistance.

During November 2009, the project identified several fragments as new candidates for a novel binding site on the peripheral surface of HIV protease. These fragments docked well against the "exo site" and in vitro studies (i.e., "wet lab" experiments in test tubes) will assess their potencies. If these wet lab experiments produce promising results, then these fragments could form the foundation for the development of "allosteric inhibitors" of HIV protease (i.e., "flexibility wedges" that can disrupt the conformational changes that HIV protease must undergo in order to function). These allosteric inhibitors could represent a totally new class of anti-AIDS compounds.

These two examples clearly explain why scientists are now routinely supported in their research by grid infrastructures. But what about business and casual users?

Although projects such as BEinGRID have reported some successful business experiments that may profit from execution in grid environments, it appears that there is not a general business case for the grid. However, recent advances in virtualization techniques, coupled with the increased Internet bandwidth now available, led in 2007 to the concept of cloud computing. The emergence of this new paradigm is mainly based on its simplicity and the affordable price for seamless access to both computational and storage resources.

Virtualization enables cloud computing, providing the ability to run legacy applications on older operating systems, creation of a single system image starting from an heterogeneous collection of machines such as those traditionally found in grid environments, and faster job migration within different virtual machines running on the same hardware. For grid and cloud computing, virtualization is the key for provisioning and fair resource allocation. From the security point of view, since virtual machines run isolated in their sandboxes, this provides an additional protection against malicious or faulty codes.

Clouds provide access to inexpensive hardware and storage resources through very simple APIs, and are based on a pay-per-use model, so that renting these resources is usually much cheaper than acquiring dedicated new ones. Moreover, people are becoming comfortable with storing their data remotely in a cloud environment. Therefore, clouds are being increasingly used by scientists, small and medium sized enterprises, and casual users.

Grids, clouds, and virtualization are exciting technologies that are going to become prominent in the next few years; we expect a wide proliferation in their use, especially clouds since these distributed computing facilities are already accessible at a reasonable cost to many potential users. We also expect grids and clouds to play an ever increasing role in the field of scientific research. It is therefore necessary a thorough understanding of principles and techniques of these fields, and the main aim of this book is to foster awareness of the essential ideas by exploring current and future developments in the area.

The idea of writing this book dates back to the highly successful Grids, Clouds and Virtualization Workshop that we organized in conjunction with the 4th International Conference on Grid and Pervasive Computing (GPC 2009) held in Geneva, 4–8 May 2009. We were contacted by Mr. Wayne Wheeler of Springer, and, after an insightful discussion, we agreed to serve as the editors for the book. Indeed, it is virtually impossible for a single person to write a book covering all of the important aspects of grids, clouds, and virtualization while maintaining the required depth, consistency, and appeal.

We invited many well-known and internationally recognized experts, asking them to contribute their expertise. The book delves into details of grids, clouds, and virtualization, guiding the reader through a collection of chapters dealing with key topics. The bibliography rather than being exhaustive, covers essential reference material. The aim is to avoid an encyclopedic approach since we believe that an attempt to cover everything will instead fail to convey any useful information to the interested readers, an audience including researchers actively involved in the field, undergraduate and graduate students, system designers and programmers, and IT policy makers.

The book may serve both as an introduction and as a technical reference. Our desire and hope is that it will be useful to many people familiarizing with the subject and will contribute to new advances in the field.

Lecce, Italy                                                                   Massimo Cafaro
                                                                               Giovanni Aloisio

# Acknowledgements

Every book requires months of preparations, and this book is no exception. We would like to express our gratitude to the contributors for their participation in this project. Without their technical expertise, patience, and efforts, this book would not have been possible.

We are also indebted with the Springer editorial team for their cooperation efforts that made this book a reality. In particular, we are deeply grateful to Mr. Wayne Wheeler, Senior Editor, for his initial proposal and continuous encouragements. We serve as the editors of this book owing to his incredible skills and energy. Special thanks must also go to Mr. Simon Rees, Senior Editorial Assistant, for his dedication, support, and punctuality.

# Contents

1   **Grids, Clouds, and Virtualization** . . . . . . . . . . . . . . . . . . . . 1
    Massimo Cafaro and Giovanni Aloisio

2   **Quality of Service for I/O Workloads in Multicore Virtualized
    Servers** . . . . . . . . . . . . . . . . . . . . . . . . . . . . . . . . . . . . 23
    J. Lakshmi and S.K. Nandy

3   **Architectures for Enhancing Grid Infrastructures with Cloud
    Computing** . . . . . . . . . . . . . . . . . . . . . . . . . . . . . . . . . . 55
    Eduardo Huedo, Rafael Moreno-Vozmediano, Rubén S. Montero, and
    Ignacio M. Llorente

4   **Scientific Workflows in the Cloud** . . . . . . . . . . . . . . . . . . . . 71
    Gideon Juve and Ewa Deelman

5   **Auspice: Automatic Service Planning in Cloud/Grid Environments** . 93
    David Chiu and Gagan Agrawal

6   **Parameter Sweep Job Submission to Clouds** . . . . . . . . . . . . . . 123
    P. Kacsuk, A. Marosi, M. Kozlovszky, S. Ács, and Z. Farkas

7   **Energy Aware Clouds** . . . . . . . . . . . . . . . . . . . . . . . . . . . 143
    Anne-Cécile Orgerie, Marcos Dias de Assunção, and Laurent Lefèvre

8   **Jungle Computing: Distributed Supercomputing Beyond Clusters,
    Grids, and Clouds** . . . . . . . . . . . . . . . . . . . . . . . . . . . . . . 167
    Frank J. Seinstra, Jason Maassen, Rob V. van Nieuwpoort, Niels Drost,
    Timo van Kessel, Ben van Werkhoven, Jacopo Urbani, Ceriel Jacobs,
    Thilo Kielmann, and Henri E. Bal

9   **Application-Level Interoperability Across Grids and Clouds** . . . . . 199
    Shantenu Jha, Andre Luckow, Andre Merzky, Miklos Erdely, and
    Saurabh Sehgal

**Glossary** . . . . . . . . . . . . . . . . . . . . . . . . . . . . . . . 231

**Index** . . . . . . . . . . . . . . . . . . . . . . . . . . . . . . . . . 233

# Contributors

**S. Ács** MTA SZTAKI, P.O. Box 63, 1518 Budapest, Hungary, acs@sztaki.hu

**Gagan Agrawal** Department of Computer Science and Engineering, The Ohio State University, Columbus, OH 43210, USA, agrawal@cse.ohio-state.edu

**Giovanni Aloisio** University of Salento, Lecce, Italy, giovanni.aloisio@unisalento.it

**Henri E. Bal** Department of Computer Science, Vrije Universiteit, De Boelelaan 1081A, 1081 HV Amsterdam, The Netherlands, bal@cs.vu.nl

**Massimo Cafaro** Dipartimento di Ingegneria dell'Innovazione, Università del Salento, Via per Monteroni, 73100 Lecce, Italy, massimo.cafaro@unisalento.it

**David Chiu** School of Engineering and Computer Science, Washington State University, Vancouver, WA 98686, USA, david.chiu@wsu.edu

**Marcos Dias de Assunção** INRIA, LIP Laboratory (UMR CNRS, INRIA, ENS, UCB), University of Lyon, 46 allée d'Italie, 69364 Lyon Cedex 07, France, marcos.dias.de.assuncao@ens-lyon.fr

**Ewa Deelman** University of Southern California, Marina del Rey, CA, USA, deelman@isi.edu

**Niels Drost** Department of Computer Science, Vrije Universiteit, De Boelelaan 1081A, 1081 HV Amsterdam, The Netherlands, niels@cs.vu.nl

**Miklos Erdely** University of Pannonia, Veszprem, Hungary, erdelyim@gmail.com

**Z. Farkas** MTA SZTAKI, P.O. Box 63, 1518 Budapest, Hungary, zfarkas@sztaki.hu

**Eduardo Huedo** Universidad Complutense de Madrid, 28040 Madrid, Spain, ehuedo@fdi.ucm.es

**Ceriel Jacobs** Department of Computer Science, Vrije Universiteit, De Boelelaan 1081A, 1081 HV Amsterdam, The Netherlands, ceriel@cs.vu.nl

**Shantenu Jha** Louisiana State University, Baton Rouge, 70803, USA,
sjha@cct.lsu.edu

**Gideon Juve** University of Southern California, Marina del Rey, CA, USA,
juve@usc.edu

**P. Kacsuk** MTA SZTAKI, P.O. Box 63, 1518 Budapest, Hungary,
kacsuk@sztaki.hu

**Thilo Kielmann** Department of Computer Science, Vrije Universiteit, De Boelelaan 1081A, 1081 HV Amsterdam, The Netherlands, kielmann@cs.vu.nl

**M. Kozlovszky** MTA SZTAKI, P.O. Box 63, 1518 Budapest, Hungary,
m.kozlovszky@sztaki.hu

**J. Lakshmi** SERC, Indian Institute of Science, Bangalore 560012, India, jlak-
shmi@serc.iisc.ernet.in

**Laurent Lefèvre** INRIA, LIP Laboratory (UMR CNRS, INRIA, ENS, UCB), University of Lyon, 46 allée d'Italie, 69364 Lyon Cedex 07, France,
laurent.lefevre@inria.fr

**Ignacio M. Llorente** Universidad Complutense de Madrid, 28040 Madrid, Spain,
llorente@dacya.ucm.es

**Andre Luckow** Louisiana State University, Baton Rouge, 70803, USA,
aluckow@cct.lsu.edu

**Jason Maassen** Department of Computer Science, Vrije Universiteit, De Boelelaan
1081A, 1081 HV Amsterdam, The Netherlands, jason@cs.vu.nl

**A. Marosi** MTA SZTAKI, P.O. Box 63, 1518 Budapest, Hungary, atisu@sztaki.hu

**Andre Merzky** Louisiana State University, Baton Rouge, 70803, USA, an-
dre@merzky.net

**Rubén S. Montero** Universidad Complutense de Madrid, 28040 Madrid, Spain,
rubensm@dacya.ucm.es

**Rafael Moreno-Vozmediano** Universidad Complutense de Madrid, 28040 Madrid,
Spain, rmoreno@dacya.ucm.es

**S.K. Nandy** SERC, Indian Institute of Science, Bangalore 560012, India,
nandy@serc.iisc.ernet.in

**Anne-Cécile Orgerie** ENS Lyon, LIP Laboratory (UMR CNRS, INRIA, ENS,
UCB), University of Lyon, 46 allée d'Italie, 69364 Lyon Cedex 07, France,
annececile.orgerie@ens-lyon.fr

**A.J. Sammes**, Centre for Forensic Computing, Cranfield University, DCMT,
Shrivenham, Swindon SN6 8LA, UK

**Saurabh Sehgal** Louisiana State University, Baton Rouge, 70803, USA

**Frank J. Seinstra** Department of Computer Science, Vrije Universiteit, De Boelelaan 1081A, 1081 HV Amsterdam, The Netherlands, fjseins@cs.vu.nl

**Jacopo Urbani** Department of Computer Science, Vrije Universiteit, De Boelelaan 1081A, 1081 HV Amsterdam, The Netherlands, jacopo@cs.vu.nl

**Timo van Kessel** Department of Computer Science, Vrije Universiteit, De Boelelaan 1081A, 1081 HV Amsterdam, The Netherlands, timo@cs.vu.nl

**Rob V. van Nieuwpoort** Department of Computer Science, Vrije Universiteit, De Boelelaan 1081A, 1081 HV Amsterdam, The Netherlands, rob@cs.vu.nl

**Ben van Werkhoven** Department of Computer Science, Vrije Universiteit, De Boelelaan 1081A, 1081 HV Amsterdam, The Netherlands, ben@cs.vu.nl

# Chapter 1
# Grids, Clouds, and Virtualization

**Massimo Cafaro and Giovanni Aloisio**

**Abstract** This chapter introduces and puts in context Grids, Clouds, and Virtualization. Grids promised to deliver computing power on demand. However, despite a decade of active research, no viable commercial grid computing provider has emerged. On the other hand, it is widely believed—especially in the Business World—that HPC will eventually become a commodity. Just as some commercial consumers of electricity have mission requirements that necessitate they generate their own power, some consumers of computational resources will continue to need to provision their own supercomputers. Clouds are a recent business-oriented development with the potential to render this eventually as rare as organizations that generate their own electricity today, even among institutions who currently consider themselves the unassailable elite of the HPC business. Finally, Virtualization is one of the key technologies enabling many different Clouds. We begin with a brief history in order to put them in context, and recall the basic principles and concepts underlying and clearly differentiating them. A thorough overview and survey of existing technologies provides the basis to delve into details as the reader progresses through the book.

## 1.1 Introduction

This chapter introduces and puts in context Grids, Clouds, and Virtualization [17]. Grids promised to deliver computing power on demand. However, despite a decade of active research, no viable commercial grid computing provider has emerged. On the other hand, it is widely believed—especially in the Business World—that HPC will eventually become a commodity. Just as some commercial consumers of electricity have mission requirements that necessitate they generate their own power,

M. Cafaro (✉) · G. Aloisio
University of Salento, Lecce, Italy
e-mail: massimo.cafaro@unisalento.it

G. Aloisio
e-mail: giovanni.aloisio@unisalento.it

M. Cafaro, G. Aloisio (eds.), *Grids, Clouds and Virtualization,*
Computer Communications and Networks,
DOI 10.1007/978-0-85729-049-6_1, © Springer-Verlag London Limited 2011

some consumers of computational resources will continue to need to provision their own supercomputers. Clouds are a recent business-oriented development with the potential to render this eventually as rare as organizations that generate their own electricity today, even among institutions who currently consider themselves the unassailable elite of the HPC business. Finally, Virtualization is one of the key technologies enabling many different Clouds. We begin with a brief history in order to put them in context, and recall the basic principles and concepts underlying and clearly differentiating them. A thorough overview and survey of existing technologies and projects provides the basis to delve into details as the reader progresses through the book.

## 1.2 A Bit of History

The history of Grids and Clouds may be traced back to the 1961 MIT Centennial, when John McCarthy, a pioneer in mathematical theory of computation and artificial intelligence and the inventor of the Lisp programming language, first exposed the idea of utility computing: "...If computers of the kind I have advocated become the computers of the future, then computing may someday be organized as a public utility just as the telephone system is a public utility...The computer utility could become the basis of a new and important industry."

Indeed, owing to the huge costs and complexity of provisioning and maintaining a data center, in the next two decades many large organizations (primarily banks) rented computing power and storage provided by mainframe computers geographically spread in the data centers of IBM and other providers. Meanwhile, mini, micro, and personal computers appeared on the market. During early 1980s, the majority of the organizations acquired affordable personal computers and workstations. This was perceived as the end of utility computing until the next decade.

In 1992, Charlie Catlett and Larry Smarr introduced the concept of metacomputing in their seminal paper [50]. The term metacomputing refers to computation on a virtual supercomputer assembled connecting together different resources like parallel supercomputers, data archives, storage systems, advanced visualization devices, and scientific instruments using high-speed networks that link together these geographically distributed resources. The main reason for doing so is because it enables new classes of applications [14, 39] previously impossible and because it is a cost-effective approach to high-performance computing. Metacomputing proved to be feasible in several experiments and testbeds, including the I-WAY experiment [19, 25] and in the Globus Gusto testbed [26].

The new applications were initially classified as follows:

- desktop supercomputing;
- smart instruments;
- collaborative environments;
- distributed supercomputing.

Desktop supercomputing included applications coupling high-end graphics capabilities with remote supercomputers and/or databases; smart instruments are scientific instruments like microscopes, telescopes, and satellites requiring supercomputing power to process the data produced in near real time. In the class of collaborative environments there were applications in which users at different locations could interact together working on a supercomputer simulation; distributed supercomputing finally was the class of applications requiring multiple supercomputers to solve problems otherwise too large or whose execution was divided on different components that could benefit from execution on different architectures.

The challenges to be faced before metacomputing could be really exploited were identified as related to the following issues:

- scaling and selection;
- unpredictable structure;
- heterogeneity;
- dynamic behavior;
- multiple administrative domains.

Interestingly (from a research perspective), these are still relevant today. Scaling is a concern, because we expect that grid and cloud environments in the future will become even larger, and resources will be selected and acquired on the basis of criteria such as connectivity, cost, security, and reliability. These resources will show different levels of heterogeneity, ranging from physical devices to system software and schedulers policies; moreover, traditional high-performance applications are developed for a single supercomputer whose features are known a priori, e.g., the latency of the interconnection network; in contrast, grid and cloud applications will run in a wide range of environments, thus making impossible to predict the structure of the computation. Another concern is related to the dynamic behavior of the computation [51], since we cannot, in general, be assured that all of the system characteristics stay the same during the course of computation, e.g., the network bandwidth and latency can widely change, and there is the possibility of both network and resource failure. Finally, since the computation will usually span resources geographically spread at multiple administrative domains, there is not a single authority in charge of the system, so that different scheduling policies and authorization mechanisms must be taken into account.

In the same years, a middleware project called Legion [33] promoted the grid object model. Legion was designed on the basis of common abstractions of the object-oriented model: everything in Legion was an object with a well-defined set of access method, including files, computing resources, storage, etc. The basic idea was to expose the grid as a single, huge virtual machine with the aim of hiding the underlying complexity to the user. The middleware proved to be useful in several experiments, including a distributed run of an ocean model and task-farmed computational chemistry simulations. However, it became immediately apparent that the majority of the people in academia were not fond of the grid object model; consequently, the attention shifted toward the use of the Globus Toolkit, which a few years later provided an alternative software stack and became quickly the de facto standard for grid computing.

The Globus Toolkit, initially presented as a metacomputing environment [24], was one of the first middleware solutions really designed to tackle the issues related to large-scale distributed computing, and its usefulness in the context of metacomputing was demonstrated in the Gusto testbed; among the distributed simulations that have been run using Globus, there was SF-Express [16], the largest computer simulation of a military battle involving at the time more that 100,000 entities. Even the EuropeanData Grid project, in charge of building a European middleware for grid computing, did so leveraging many of the Globus components. These efforts led to the gLite middleware [40], which is going to be used for the analysis of experimental results of the Large Hadron Collider, the particle accelerator at CERN that recently started operations in Geneva. The middleware was also actively tested in the context of the EGEE project [41], which provided the world largest computational grid testbed to date. Another European project developed a different middleware, Unicore [21], targeting mainly HPC resources. A similar endeavor, devoted to the implementation of grid middleware for high-performance computing, took place in Japan starting in 2003 with the National Research Grid Initiative (NAREGI) [43] and culminating in 2008 with the release of the middleware.

There was a pressure to grid-enable many existing legacy products. Among the schedulers, we recall here Condor [52], Nimrod [10], Sun Grid Engine [30], and Platform Computing LSF [56]. The Condor project begun in 1988 and is therefore one of the earliest cycle scavenging software available. Designed for loosely coupled jobs, and to cope with failures, it was ported to the grid through a Globus extension aptly named Condor-G. The Nimrod system was designed to manage large parameter sweep jobs, in which the same executable was run each time with a different input. Since the jobs are independent, a grid is clearly a good fit for this situation. The Nimrod-G system, extended once again through the Globus Toolkit, was also one of the earliest projects in which the concepts of grid economy appeared. In addition to several other criteria, the user could also take into account the cost of the whole simulation when submitting a job. The system was able to charge the CPU cycles distinguishing HPC resources from traditional, off-the-shelf ones. Sun Grid Engine, an open source project led by Sun, started in 2000. The software was originally based on the Gridware Codine (COmputing in DIstributed Network Environments) scheduler. It was grid-enabled and is now being cloud-enabled. Finally, LSF was one the first commercial products offering support for grid environments through its extension named LSF MultiCluster.

The business and commercial world recognized the impact and the potential of grid computing, whose usefulness was demonstrated in hundreds of projects. However, almost all of the projects were driven by people involved in academia and targeting mainly scientific applications. Very few projects, notably the BEinGrid [20] one, were in charge of demonstrating the use of grid computing for business oriented goals. BEinGrid, with its sample of 25 business experiments, was also successful in the implementation and deployment of Grid solutions in industrial key sectors. Nonetheless, grids are still not appealing to the business world, and this is reflected in the lack of current commercial grid solutions. Platform Computing provided years ago a version of the Globus Toolkit which is now unsupported and not for sale. Legion was initially sold by Applied Meta, then by Avaki Corporation which was acquired in 2005 by Sybase, Inc. What remains of the software is now used by the company's customers to share enterprise data. Entropia, Inc., a company founded in 1997, sold distributed computing software for CPU scavenging until 2004.

The idea of harnessing idle CPUs worldwide was first exploited by the SETI@home project [11], launched in 1996. Soon a number of related projects including GIMPS [44], FightAIDS@Home [47], Folding@home [13] arose. GIMPS, the Great Internet Mersenne Prime Search, was started in January 1996 to discover new world-record-size Mersenne primes. A Mersenne prime is a prime of the form $2^P - 1$. The first Mersenne primes are $3, 7, 31, 127$ (corresponding to $P = 2, 3, 5, 7$). There are only 47 known Mersenne primes, and GIMPS has found 13 of the 47 Mersenne primes ever found during its 13-year history. FightAIDS@Home uses distributed computing exploiting idle resources to accelerate research into new drug therapies for HIV, the virus that causes AIDS; FightAIDS@Home made history in September 2000 when it became the first biomedical Internet-based grid computing project. Proteins, in order to carry out their function, must take on a particular shape, also known as a fold. One of the Folding@home goals is to simulate protein folding in order to understand how proteins fold so quickly and reliably, and to learn about what happens when this process goes awry (when proteins misfold).

The initial release of the Apple Xgrid [35, 36] technology was also designed for independent, embarrassingly parallel jobs, and was later extended to support tightly coupled parallel MPI jobs. A key requirement of Xgrid was simplicity: everyone is able to setup and use an hoc grid using Xgrid, not just scientists. However, only Apple resources running the Mac OS X operating system may belong to this grid, although third-party software (an Xgrid Linux agent and a Java-based one) not supported or endorsed by Apple allows deploying heterogeneous grids.

## 1.3 Grids

It is interesting to begin by reviewing some of most important definitions of grid computing given during the course of research and development of the field. The earliest definition given in 1998 by Foster and Kesselman [37] is focused around on-demand access to computing, data, and services: a computational grid is a hardware and software infrastructure that provides dependable, consistent, pervasive, and inexpensive access to high-end computational capabilities. Two years later, the definition was changed to reflect the fact that grid computing is concerned with coordinated resource sharing and problem solving in dynamic, multi-institutional virtual organizations. In the latest definitions of Foster [23, 27], a grid is described respectively as an infrastructure for coordinated resource sharing and problem solving in dynamic, multi-institutional virtual organizations and as a system that coordinates resources that are not subject to centralized control, using standard, open, general-purpose protocols and interfaces to deliver nontrivial qualities of service. We will argue later, when comparing grids and clouds, that delivering nontrivial qualities of service is very difficult using a distributed approach instead of a centralized one.

Grids are certainly an important computing paradigm for science, and there are many different scientific communities facing large-scale problems simply too big to be faced even on a single, powerful parallel supercomputer. After more than a decade of active research and development in the field, all of the people expected grids to become a commercial service; however, this has not happened yet. Therefore, we ask ourselves: is there a general business case for grids?

To answer this question, we recall that it is now clear that grid computing benefits specific classes of applications and that the technology itself, while powerful, is probably not yet simple enough to be released commercially. This is reflected in the following sentence of Frank Gillette, Forester: "None of us have figured out a simple way to talk about (grid) . . . because it isn't simple." And, when something takes an hour or more to be explained, it certainly cannot be sold easily. Therefore, application providers may prefer to integrate the required middleware into their applications or take advantage of existing Platform and DataSynapse solutions. The issue raised by the complexity of grids proves to be a formidable barrier that we must overcome if we want grids to become a fact of life that everyone uses and nobody notices. This has been recently remarked in a position paper by Fox and Pierce [29] in which the authors discuss the fact that "Grids meet Too much Computing, Too much Data and never Too much Simplicity."

On the other hand, early testbeds and recent grids deployment have clearly shown the potential of grid computing. In particular, loosely coupled parallel applications, parameter sweep studies and applications described by a workflow assembling and orchestrating several different components are viable candidates for grid computing. Example applications include those from the High Energy Physics (HEP) community, falling in the parameter sweep class, and those from the bioinformatics community, falling in the workflow class. Typical loosely coupled parallel applications are exemplified by climate simulations, in which the initial functional decomposition of the problem is well suited for execution on grids linking together distributed supercomputers; the blocks identified by the initial decomposition (atmosphere, land, sea, ice, etc.) are then run on those supercomputers and exchange data infrequently.

When the vendors realized that this kind of applications were far beyond the common needs of the majority of the people, the attention shifted to the emerging field of cloud computing. Here, in contrast to grid computing, the emphasis was put on simplicity as the ingredient to make clouds easily accessible.

## 1.4 Clouds

It is rather difficult to precisely define what clouds and cloud computing are, especially taking into account the many possible different uses. We recall here that the same also happened in the context of grids. The concept as we all know it today was introduced in late 2007 [54]. Among the many definitions that were given since then, we report the following ones:

- Gartner: Cloud computing is a style of computing where massively scalable IT-related capabilities are provided "as a service" across the Internet to multiple external customers;
- Forrester: A pool of abstracted, highly scalable, and managed infrastructure capable of hosting end-customer applications and billed by consumption;
- IBM: An emerging computing paradigm where data and services reside in massively scalable data centers and can be ubiquitously accessed from any connected devices over the internet.

How clouds are actually perceived by the people appears to be much broader in scope. The following five definitions [48] expose several perspectives on this subject.

1. Cloud computing refers (for many) to a variety of services available over the Internet that deliver compute functionality on the service provider's infrastructure (e.g., Google Apps, Amazon EC2, or Salesforce.com). A cloud computing environment may actually be hosted on either a grid or utility computing environment, but that does not matter to a service user;
2. • Cloud computing = Grid computing. The workload is sent to the IT infrastructure that consists of dispatching masters and working slave nodes. The masters control resource distributions to the workload (how many slaves run the parallelized workload). This is transparent to the client, who only sees that workload has been dispatched to the cloud/grid and results are returned to it. The slaves may or may not be virtual hosts;
   • Cloud computing = Software-as-Service. This is the Google Apps model, where apps are located "in the cloud," i.e., somewhere in the Web;
   • Cloud computing = Platform-as-Service. This is the Amazon EC2 et al. model, where an external entity maintains the IT infrastructure (masters/slaves), and the client buys time/resources on this infrastructure. This is "in the cloud" in so much that it is across the Web, outside of the organization that is leasing time off it;
3. The cloud simply refers to the move from local to service on the Web. From storing files locally to storing them in secure scalable environments. From doing apps that are limited to GB spaces to now apps that have no upper boundary, from using Microsoft Office to using a Web-based office. Somewhere in 2005–2008 storage online got cheaper and more secure than storing locally or on your own server. This is the cloud. It encompasses grid computing, larger databases like Bigtable, caching, always accessible, failover, redundant, scalable, and all sorts of things. Think of it as a further move into the Internet. It also has large implications for such battles as static vs. dynamic, RDBMS vs. BigTable and flat data views. The whole structure of business that relies on IT infrastructure will change, programmers will drive the cloud, and there will be lots of rich programmers at the end. It is like the move from mainframe to personal computers. Now you have a personal space in the clouds;
4. Grid and Cloud are not exclusive of each other... Our customers view it this way: Cloud is pay for usage (i.e., you do not necessarily own the resources), and

Grid is how to schedule the work, regardless where you run it. You can use a cloud without a grid and a grid without a cloud. Or you can use a grid on a cloud;
5. I typically break up the idea of cloud computing into three camps:

- Enablers. These are companies that enable the underlying infrastructures or the basic building blocks. These companies are typically focused on data center automation and or server virtualization (VMware/EMC, Citrix, BladeLogic, RedHat, Intel, Sun, IBM, Enomalism, etc.);
- Providers (Amazon Web Services, Rackspace, Google, Microsoft). The ones with the budgets and know-how to build out global computing environments costing millions or even billions of dollars. Cloud providers typically offer their infrastructure or platform. Frequently, these As-a-Service offerings are billed and consumed on a utility basis;
- Consumers. On the other side of the spectrum, I see the consumers companies that build or improve their Web applications on top of existing clouds of computing capacity without the need to invest in data centers or any physical infrastructure. Often these two groups can be one in the same such as Amazon (SQS, SDB, etc.), Google (Apps), and Salesforce (Force). But they can also be new startups that provide tools and services that sit on top of the cloud (Cloud management).

Cloud consumers can be a fairly broad group including just about any application that is provided via a Web-based service like a Webmail, blogs, social network, etc. Cloud computing from the consumer point of view is becoming the only way you build, host, and deploy a scalable Web application.

On the other hand, the main findings of the Cloud BoF held at OGF22, Cambridge, MA, on 27 Feb 2008 were the following ones:

- Clouds are "Virtual Clusters" ("Virtual Grids") of possibly "Virtual Machines";
- They may cross administrative domains or may "just be a single cluster"; the user cannot and does not want to know;
- Clouds support access (lease of) computer instances;
- Instances accept data and job descriptions (code) and return results that are data and status flags;
- Each Cloud is a "Narrow" (perhaps internally proprietary) Grid;
- When does Cloud concept work:
  - Parameter searches, LHC style data analysis, ...
  - Common case (most likely success case for clouds) versus corner case?
- Clouds can be built from Grids;
- Grids can be built from Clouds;
- Geoffrey Fox: difficult to compare grids and clouds because neither term has an agreed definition;
- Unlike grids, clouds expose a simple, high-level interface;
- There are numerous technical issues:
  - performance overhead, cost, security, computing model, data-compute affinity, schedulers and QoS, link clouds (e.g., compute-data), ...;

– What happens when a cloud goes down? What about interoperability of clouds? Standardization? Is it just another service?

A recent report [12] present an in-depth view of cloud computing. In what follows we will try to make things easier to understand, summarizing our perspective. In order to do so, we begin discussing the main features and characteristics of clouds as currently deployed and made available to the people. The main characteristic of cloud computing certainly is its focus on virtualization. Clouds are succeeding owing to the fact that the underlying infrastructure and physical location are fully transparent to the user.

Beyond that, clouds also exhibit excellent scalability allowing users to run increasingly complex applications and breaking the overall workload into manageable pieces served by the easily expandable cloud infrastructure. This flexibility is a key ingredient, and it is very appealing to the users. Clouds can adapt dynamically to both consumer and commercial workloads, providing efficient access through a Service Oriented Architecture to a computing infrastructure delivering dynamic provisioning of shared compute resources.

An attempt to provide a comprehensive comparison of grids and clouds is of course available [28]. However, we now compare and contrast grids and clouds, in order to highlight what we think are key differences. Grids are based on open standards; the standardization process happens in organizations such as the Open Grid Forum, OASIS, etc. Clouds, in contrast, do not provide standardized interfaces. Especially commercial clouds solutions are based on proprietary protocols, which are not disclosed to the scientific community. A related aspect is that of interoperability. While in grid environments interoperability has become increasingly important, and many efforts are devoted to this topic [15], clouds are not interoperable and will not be in the short-term period. Vendors have no interest at the moment to provide interoperability among their cloud infrastructures.

Almost all of the currently deployed grids have been publicly funded and operated. This is of course a very slow process, when compared to clouds that are instead privately funded and operated. Many companies have already invested and continue to invest a huge amount of money to develop cloud technologies; however, considering the limited amount of funding available to scientists, we must remark the excellent results in the field of grid computing.

Examining how grids are operated, it is easy to see that, since the beginning, there was a design requirement to build grid infrastructure tying together distributed administrative domains, possibly geographically spread. On the contrary, clouds are managed by a single company/entity/administrative domain. Everything is centralized, the infrastructure is hosted in a huge centre, and only the clients are actually geographically distributed. The architecture is basically client–server.

We note here that, despite its simplicity, the client–server architecture is still the most widely used in the commercial world, owing to the fact that it is very hard to beat hosted/managed services with regard to performance and resiliency: these services are currently geographically replicated and hugely provisioned and can guarantee/meet a specific Service Level Agreement. Highly distributed architectures, including peer-to-peer systems, while in principle are theoretically more resistant to threats such as Denial of Service attacks etc., still do not provide the same performance guarantee. For instance, a P2P Google service is deemed to be impossible [42] (although Google uses internally their own P2P system, based on the concept of MapReduce [18]), and people who do it for business, today do not do it peer-to-peer, with the notable exception of the Skype service [49].

Coming to the main purpose of grid and cloud systems, it is immediately evident that for grid systems, the main raison d'etre is resource sharing. Cloud systems are instead intended to provide seamless access to a huge, scalable infrastructure. Given the different purpose they serve, it comes to no surprise that these systems target different classes of applications. As we have already noted, grids are well suited for scientific research and for the needs of high-end users. Clouds are mainly used today for data analysis, information processing, and data mining.

We conclude this section recalling that Many Task Computing (MTC) and High Throughput Computing (HTC) service providers and resource service providers can benefit from the economies of scale that clouds can deliver [53], making this kind of infrastructure appealing to the different stakeholders involved in.

## 1.5 Virtualization

The Virtual Machine (VM) concept [45] dates back to the 1960s; it was introduced by IBM as a mean to provide concurrent, interactive access to their mainframe computers. A VM was an instance of the physical machine and gave users the illusion of accessing the physical machine directly. VMs were used to transparently enable time-sharing and resource-sharing on the (at the time) highly expensive hardware. Each VM was a fully protected and isolated copy of the underlying system. Virtualization was thus used to reduce the hardware costs on one side and to improve the overall productivity by letting many more users work on it simultaneously. However, during the course of years, the hardware got cheaper and simultaneously multiprocessing operating systems emerged. As a consequence, VMs were almost extinct in 1970s and 1980s, but the emergence of wide varieties of PC-based hardware and operating systems in 1990s revived virtualization ideas.

Virtualization technologies represent a key enabler of cloud computing, along with the recent advent of Web 2.0 and the increased bandwidth availability on the Internet. The most prominent feature is the ability to install multiple OS on different virtual machines on the same physical machine. In turn, this provides the additional benefits of overall cost reduction owing to the use of less hardware and consequently less power. As a useful side effect, we also note here that virtualization generally leads to increased machine utilization. The main aim of virtualization technologies is to hide the physical characteristics of computing resources from the way in which other systems, applications, or end users interact with those resources. In this book, Lakshmi et al. [38] propose an end-to-end system virtualization architecture and thoroughly analyze it.

Among the many benefits provided by virtualization, we recall the ability to run legacy applications requiring an older platform and/or OS, the possibility of creating a single system image starting from an heterogeneous collection of machines such as those traditionally found in grid environments, and faster job migration within different virtual machines running on the same hardware. For grid and cloud computing, virtualization is the key for provisioning and fair resource allocation. From the security point of view, since virtual machines run isolated in their sandboxes, this provides an additional protection against malicious or faulty codes.

Besides computing, storage may be virtualized too. Through the aggregation of multiple smaller storage devices characterized by attributes such as performance, availability, capacity and cost/capacity, it becomes possible to present them as one or more virtual storage devices exhibiting better performance, availability, capacity, and cost/capacity properties. In turn, this clearly enhances the overall manageability of storage and provides better sharing of storage resources.

A virtualization layer provides the required infrastructural support exploiting lower-level hardware resources in order to create multiple independent virtual machines that are isolated from each other. This layer, traditionally called Virtual Machine Monitor (VMM), usually sits on top of the hardware and below the operating system. Virtualization as an abstraction can take place at several different levels, including instruction set level, Hardware Abstraction Layer (HAL), OS level (system call interface), user-level library interface, and the application level.

Virtualizing the instruction set requires emulation of the instruction set, i.e., interpreting the instructions in software. A notable example is Rosetta [9], which is a lightweight dynamic translator for Mac OS X distributed by Apple. Its purpose is to enable legacy applications compiled for the PowerPC family of processors to run on current Apple systems that use Intel processors. Other examples include Bochs [1], QEMU [2], and BIRD [46].

HAL level virtual machines are based on abstractions lying between a real machine and an emulator; in this case a virtual machine is an environment created and managed by a VMM. While emulator's functionalities provide a complete layer between the operating system or applications and the hardware, a VMM is in charge of managing one or more VMs, and each VM in turn provides facilities to an OS or application as if it is run in a normal environment, directly on the hardware. Examples include VMware [3], Denali [55], Xen [4], Parallels [5], and Plex86 [6].

Abstracting virtualization at the OS level requires providing a virtualized system call interface; this involves working on top of the OS or at least as an OS module. Owing to the fact that system calls are the only way of communication from user-space processes to kernel-space, the virtualization software can control user-space processes by managing this system interface. Additionally, besides system's library, usually applications also link code provided by third-party user-level libraries. Virtualizing this library interface can be done by controlling the communication link between the applications and the rest of the system through well-defined API hooks. Therefore, it is possible to expose a different implementation using the same set of APIs. As an example, WINE HQ [7] and CrossOver [8] support running Windows applications respectively on Unix and Mac OS X.

Virtualization at the application is not based on the insertion of a virtualization layer in the middle. Instead, this kind of abstraction is implemented as an application that eventually creates a virtual machine. The created VM could range in complexity from a simple language interpreter to a very complex Java Virtual machine (JVM).

All of these virtualization technologies are based on approaches differing significantly when evaluated with regard to several metrics such as performance, flexibility, simplicity, resource consumption, and scalability. Therefore, it is important to understand their usage scenarios as well. Instruction set emulators are often characterized by very high latencies that make them impractical to use on a regular basis (with the notable exception of Rosetta). Their main goal is to provide an environment especially tailored to debugging and learning purposes, since every component is implemented in software and is fully under user's control.

Commercial virtual machines operating at HAL level, besides offering extremely low latencies, also give the flexibility of using different OSes or different versions of the same OS on the same machine; these VMs present a complete machine interface, but of course this demands a much higher amount of resources. OS level virtual machines are useful in those cases in which this kind of flexibility is not required, because resource requirements are much lower, performance better, and manipulations (e.g., creation) faster. It is worth noting here that the attained level of isolation is lower, since all of the VMs use the same kernel and can potentially affect the whole system.

Finally, library-level virtualization technologies prove to be extremely lightweight. On the contrary, application-level virtualization suffers from extra overhead being in the application-level. The latter technology is increasingly applied in mobile computing and as a building block for trusted computing infrastructures.

## 1.6 Technologies

In this section, we review the most important technologies available. We recall here the following:

- Amazon Elastic Compute Cloud (EC2);
- Google App Engine;
- Hadoop;
- HP Labs Enterprise Cloud Assure and Cloud Software Platform (Cirious);
- IBM Smart Business cloud solutions;
- Sun Cloud;
- Oracle On Demand and Platform for SaaS;
- SGI Cyclone;
- Microsoft Azure.

The Amazon cloud is probably the most used cloud computing environment. It works as follows. You start by creating an Amazon Machine Image (AMI) containing all your software, including your operating system and associated configuration settings, applications, libraries, etc. Amazon provides all of the necessary tools to create and package your AMI. Then, you upload this AMI to the cloud through the Amazon S3 (Amazon Simple Storage Service) service. This provides reliable, secure access to your AMI. In order to run your applications, you need to register your AMI with Amazon EC2. This step allows verifying that your AMI has been uploaded correctly and uniquely identifies it with an identifier called AMI ID. Using the AMI ID and the Amazon EC2 web service APIs, it is easy to run, monitor, and terminate as many instances of the AMI as required. Amazon provides command line tools and Java libraries, and you may also take advantage of SOAP or Query based APIs. Amazon also lets you launch preconfigured AMIs provided by Amazon or shared by another user. While AMI instances are running, you are billed for the computing and network resources that they consume.

Among the applications that have actually been run on EC2, we recall here web hosting, parallel processing, graphics rendering, financial modeling, web crawling, and genomics analysis. To be fair, while parallel applications can be run on the Amazon cloud, performances are of course far from the current result you can achieve on a dedicated supercomputing infrastructure, owing to less powerful CPUs and basic interconnection networks with very high latency and low bandwidth when compared to the high-end interconnects traditionally available on supercomputers. For instance, a recent work [22] is related to deploying and running a coupled climate simulation on EC2. The results obtained are comparable to running the simulation on a low-cost beowulf cluster assembled using commodity, off-the-shelf hardware.

The Google App Engine exploits the underlying Google infrastructure, the same that was built to bring access to their web search engine and Gmail to people worldwide. Leveraging a powerful network of data centers, this infrastructure is aimed at scalable web applications built using Python and, more recently, Java. Developers are assigned a free quota of 500 MB of storage and enough CPU and network bandwidth to sustain around 5 million page views per month for a typical app; it is possible to purchase additional storage and bandwidth to go beyond the limits. It is worth noting here that Google provides proprietary APIs to take advantage of the infrastructure, so that writing an application is considerably easier, but at the same time, once a development team selects Google App Engine as the environment of choice for their web application, they remain locked to the platform, because a move will require significant code rewrites. The competing Amazon EC2 service is different, in that it allows developers to create virtual server instances and leaves them the option of moving their applications easily to other machines.

Hadoop [34] is an open-source Apache Software Foundation project sponsored by Yahoo! whose intent is to reproduce the proprietary software infrastructure developed by Google. It provides a parallel programming model, MapReduce, heavily exploited by Google, a distributed file system, a parallel database, a data collection system for managing large distributed systems, a data warehouse infrastructure that provides data summarization and ad hoc querying, a high-level data-flow language and execution framework for parallel computation, and a high-performance coordination service for distributed applications.

The software is widely used by many companies and researchers. As an example of usage, we recall here the New York Times task of generating about 11 millions of PDF files related to the public domain articles from 1851–1922 by gluing together multiple TIFF images per article [31]. Derek Gottfried uploaded 4 TB of data to the

Amazon EC2 infrastructure using the S3 service, wrote the software to pull all the parts that make up an article out of S3, generate a PDF from them, and store the PDF back in S3 using open-source Java components, and finally he deployed Hadoop on a cluster of EC2 machines. Overall, the whole task took less than 24 hours using 100 instances on Amazon EC2 and generated additional 1.5 TB of data to be stored in S3.

Regarding the MapReduce programming model, it is a data-flow model between services where services can do useful document-oriented data parallel applications including reductions; using Hadoop, the decomposition of services onto cluster in a clouds is fully automated. The key idea is inspired by some primitives of the LISP programming language. Both the input and the output are key/value pairs, and developers needs to implement two functions, map and reduce:

- $map(in\_key, in\_value) \rightarrow list(out\_key, intermediate\_value)$;
- $reduce(out\_key, list(intermediate\_value)) \rightarrow list(out\_value)$.

The former processes an input key/value pair, producing a set of intermediate pairs. The latter is in charge of combining all of the intermediate values related to a particular key, outputting a set of merged output values (usually just one). MapReduce is often explained illustrating a possible solution to the problem of counting the number of occurrences of each word in a large collection of documents. The following pseudocode refers to the functions that needs to be implemented.

```
map(String input_key, String input_value):
    // input_key: document name
    // input_value: document contents
    for each word w in input_value:
      EmitIntermediate(w, "1");

  reduce(String output_key,
         Iterator intermediate_values):
    // output_key: a word
    // output_values: a list of counts
    int result = 0;
    for each v in intermediate_values:
      result += ParseInt(v);
    Emit(AsString(result));
```

The map function emits in output each word together with an associated count of occurrences (in this simple example just one). The reduce function provides the required result by summing all of the counts emitted for a specific word. MapReduce implementations (Google App Engine and Hadoop for instance) then automatically parallelize and execute the program on a large cluster of commodity machines. The runtime system takes care of the details of partitioning the input data, scheduling the program's execution across a set of machines, handling machine failures, and managing required intermachine communication. While MapReduce certainly limits the scope of parallel applications that can be developed to a particular type of interaction, it allows programmers with no previous experience with parallel and distributed computing to easily utilize the resources of a large distributed system. A typical MapReduce computation processes many terabytes of data on hundreds or thousands of machines. Programmers find the model easy to use (and it certainly is when compared to more advanced interactions among tasks), and more than 100K MapReduce jobs are routinely executed on Google's and Amazon's clusters every day.

## 1.7 The Economics

Grid Computing leverages the techniques of clustering where multiple independent clusters act like a grid due to their nature of not being located in a single domain. This leads to the data-compute affinity problem: because of the distributed nature of the Grid, the computational nodes could be located anywhere in the world. It is fine having all that CPU power available, but the data on which the CPU performs its operations could be thousands of miles away, causing a delay (latency) between data fetch and execution. CPUs need to be fed with different volumes of data depending on the tasks they are processing. Running a data intensive process with disparate data sources can create an I/O bottleneck causing the CPU to run inefficiently and affecting economic viability. This also applies to clouds but, owing to the centralized nature of cloud environments and to the requirements of cloud applications which usually move a relatively scarce volume of data (with some exceptions), the problem is far less critical. Storage management, security provisioning, and data movement are the nuts to be cracked in order for grids to succeed; more important than these technical limitations is becoming the lack of business buy in.

The nature of Grid/Cloud computing means that a business has to migrate its application and data to a third-party solution. This creates huge barriers to the uptake, at least initially. Financial business institutions need to know that grid vendors understand their business, not just the portfolio of applications they ran and the infrastructure they ran upon. This is critical to them. They need to know that whoever supports their systems knows exactly what the effect of any change could potentially make to their shareholders. The other bridge that has to be crossed is that of data security and confidentiality. For many businesses, their data is the most sensitive, business critical thing they possess. To hand this over to a third-party is simply not going to happen. Banks are happy to outsource part of their services but want to be in control of the hardware and software—basically using the outsourcer as an agency for staff.

Jim Gray (Turing Award in 1998, lost at sea in 2007) published in 2003 an interesting paper on Distributed Computing Economics [32]. As stated in this 2003 paper, "Computing economics are changing. Today there is rough price parity between (1) one database access, (2) ten bytes of network traffic, (3) 100,000 instructions, (4) 10 bytes of disk storage, and (5) a megabyte of disk bandwidth. This has implications for how one structures Internet-scale distributed computing: one puts computing as close to the data as possible in order to avoid expensive network traffic". This is exactly the data-compute affinity problem that plagues Grids.

The recurrent theme of this analysis is that On-Demand computing is only economical for very CPU-intensive (100,000 instructions per byte or a CPU-day-per gigabyte of network traffic) applications. Preprovisioned computing is likely to be more economical for most applications, especially data-intensive ones. If Internet access prices drop faster than Moore's law, the analysis fails. If instead Internet access prices drop slower than Moore's law, the analysis becomes even stronger.

What are the economic issues of moving a task from one computer to another or from one place to another? A computation task has four characteristic demands:

- Networking—delivering questions and answers,
- Computation—transforming information to produce new information,
- Database Access—access to reference information needed by the computation,
- Database Storage—long-term storage of information (needed for later access).

The ratios among these quantities and their relative costs are pivotal: it is fine to send a GB over the network if it saves years of computation, but it is not economic to send a kilobyte question if the answer could be computed locally in a second.

### 1.7.1 The Economics in 2003

To make the economics tangible, take the following baseline hardware parameters:

1. 2-GHz CPU with 2-GB RAM (cabinet and networking);
2. 200-GB disk with 100 accesses/second and 50-MB/s transfer;
3. 1-Gbps Ethernet port-pair;
4. 1-Mbps WAN link.

From this Gray concludes that one dollar equates to

- 1 GB sent over the WAN;
- 10 Tops (tera cpu operations);
- 8 hours of CPU time;
- 1 GB disk space;
- 10 M database accesses;
- 10 TB of disk bandwidth;
- 10 TB of LAN bandwidth.

Let us now think about some of the implications and consequences. Beowulf networking is 10,000× cheaper than WAN networking, and a factor of 105 matters. Moreover, the cheapest and fastest way to move a Terabyte cross country is sneakernet (the transfer of electronic information, especially computer files, by physically carrying removable media such as magnetic tape, compact discs, DVDs, USB flash drives, or external drives from one computer to another): 24 hours = 4 MB/s, 50$ shipping vs 1,000$ WAN cost; a few real examples follow. Google has reportedly used a sneakernet to transport datasets too large for current computer networks, up to 120 TB in size. The SETI@home project uses a sneakernet to overcome bandwidth limitations: data recorded by the radio telescope in Arecibo, Puerto Rico, is stored on magnetic tapes which are then shipped to Berkeley, California, for processing. In 2005, Jim Gray reported sending hard drives and even "metal boxes with processors" to transport large amounts of data by postal mail.

The conclusions of Gray (http://research.microsoft.com/en-us/um/people/gray/talks/WebServices_Grid.ppt) are that "To the extent that computational grid is like Seti@Home or ZetaNet or Folding@home or... it is a great thing; The extent that the computational grid is MPI or data analysis, it fails on economic grounds: move the programs to the data, not the data to the programs. The Internet is NOT the cpu backplane. The USG should not hide this economic fact from the academic/scientific research community".

### 1.7.2 The Economics in 2010

It is worth recalling here that when Jim Gray published his paper, the fastest Supercomputer operated at a speed of 36 TFLOPS; the fastest supercomputer available today provides computing power in excess of 1.7 PFLOPS, and a new IBM Blue Gene/Q is planned for 2010–2012, which will operate at 10 PFLOPS, practically outstripping Moore's law by a factor of 10. Internet access prices have fallen, and bandwidth has increased since 2003, but more slowly than processing power. Therefore, the economics are even worse than in 2003, and it follows that, at the moment, there are too many issues and costs with network traffic and data movements to allow Clouds to happen for all kinds of customers. The majority of routine tasks, which are not processor intensive and time critical, are the most likely candidates to be migrated to cloud computing, even though these may be the least economical to be ported to that architecture. Among the many processor intensive applications, a few selected applications such as image rendering or finite modelling appear to be good candidate applications.

## 1.8 Applications

We now briefly consider what type of applications are best suited for execution on grids and clouds respectively. The reader should take into account the issues we discussed regarding the movement of both input and output data, especially regarding large datasets. Traditional sequential batch jobs can be run without problems on both grids and clouds.

Parallel jobs can be run without issues on a single parallel supercomputer or cluster belonging to a grid; running them on clouds requires an initial effort to properly setup the environment and is likely to achieve performances similar to those on a self-made, assembled beowulf cluster using a low-cost interconnection network and off-the-shelf components.

Parameter sweep applications, in which the same executable is run each time using a different set of input files, are good candidates for execution in both grids and clouds environments: in general, loosely coupled applications with infrequent or absent interprocess communications perform well.

Finally, complex applications with data and compute dependencies are usually described as workflow applications and characterized by a graph whose vertices represent applications and whose arcs model dependencies among the vertices. When all of the nodes in the graph represent sequential batch jobs, a workflow can be run on clouds (we already noted that parallel applications are not well suited to clouds). Conversely, if some of the nodes requires parallel processing, grids are more appropriate.

## 1.9 Conclusions

This chapter introduced the concepts of grids, clouds, and virtualization. We have discussed the key role of grids in advancing the state of the art in distributed computing, and how the interest for these technologies gave rise to a plethora of projects, aimed both at developing the needed middleware stack and exploiting the new paradigm to accelerate science discovery. We have seen that despite a decade of active research, no viable commercial grid computing provider has emerged, and have investigated the reasons behind. The main problems seem to be related to the underlying complexity of deploying and managing a grid, even though there are scientific applications that may actually benefit from grid computing (in particular, complex applications requiring a distributed workflow model of execution). The other problem, data-compute affinity, affects both grids and clouds, even though in the cloud case the volume of data to be transferred is usually relatively scarce. We have then examined the emerging cloud computing paradigm, comparing and contrasting it with grids from many perspectives. Clouds build on virtualization technologies which act as one of the key enablers, and are well suited to both loosely coupled applications with infrequent or absent interprocess communications (such as parameter sweep studies) and to MapReduce-based applications. Our review of virtualization/cloud technologies puts in context cloud computing environments. Finally, we have briefly discussed the economics behind grids and clouds, laying out the foundations and providing fertile ground for the next chapters of this book.

**Acknowledgements**    The authors with to thank Martin Walker for the insightful discussions on grids and clouds in the context of the SEPAC grid project; part of the materials of this chapter are based on one of his many interesting presentations.

## References

1. URL http://bochs.sourceforge.net
2. URL http://wiki.qemu.org

3. URL http://www.vmware.com
4. URL http://xen.org
5. URL http://www.parallels.com
6. URL http://plex86.sourceforge.net
7. URL http://www.winehq.org
8. URL http://www.codeweavers.com/products/cxmac
9. URL     http://developer.apple.com/legacy/mac/library/documentation/MacOSX/Conceptual/ universal_binary/universal_binary.pdf
10. Abramson, D., Buyya, R., Giddy, J.: A computational economy for grid computing and its implementation in the Nimrod-G resource broker. Future Gener. Comput. Syst. **18**(8), 1061–1074 (2002). doi:10.1016/S0167-739X(02)00085-7
11. Anderson, D.P., Cobb, J., Korpela, E., Lebofsky, M., Werthimer, D.: Seti@home: an experiment in public-resource computing. Commun. ACM **45**(11), 56–61 (2002). doi:10.1145/581571.581573
12. Armbrust, M., Fox, A., Griffith, R., Joseph, A.D., Katz, R.H., Konwinski, A., Lee, G., Patterson, D.A., Rabkin, A., Stoica, I., Zaharia, M.: Above the clouds: a Berkeley view of cloud computing. Tech. Rep. UCB/EECS-2009-28, EECS Department, University of California, Berkeley (2009). URL http://www.eecs.berkeley.edu/Pubs/TechRpts/2009/EECS-2009-28.html
13. Beberg, A.L., Ensign, D.L., Jayachandran, G., Khaliq, S., Pande, V.S.: Folding@home: Lessons from eight years of volunteer distributed computing. In: IPDPS '09: Proceedings of the 2009 IEEE International Symposium on Parallel & Distributed Processing, pp. 1–8. IEEE Computer Society, Washington (2009). doi:10.1109/IPDPS.2009.5160922
14. Benger, W., Foster, I.T., Novotny, J., Seidel, E., Shalf, J., Smith, W., Walker, P.: Numerical relativity in a distributed environment. In: Proceedings of the Ninth SIAM Conference on Parallel Processing for Scientific Computing. SIAM, Philadelphia (1999)
15. Boardman, R., Crouch, S., Mills, H., Newhouse, S., Papay, J.: Towards grid interoperability. In: All Hands Meeting 2007, OMII-UK Workshop (2007)
16. Brunett, S., Davis, D., Gottschalk, T., Messina, P., Kesselman, C.: Implementing distributed synthetic forces simulations in metacomputing environments. In: HCW '98: Proceedings of the Seventh Heterogeneous Computing Workshop, p. 29. IEEE Computer Society, Los Alamitos (1998)
17. Cafaro, M., Aloisio, G. (eds.): Grids, Clouds and Virtualization. Springer, Berlin (2010)
18. Dean, J., Ghemawat, S.: Mapreduce: simplified data processing on large clusters. Commun. ACM **51**(1), 107–113 (2008). doi:10.1145/1327452.1327492
19. Defanti, T., Foster, I., Papka, M.E., Kuhfuss, T., Stevens, R., Stevens, R.: Overview of the I-EWAY: wide area visual supercomputing. Int. J. Supercomput. Appl. **10**(2), 123–130 (1996)
20. Dimitrakos Theoand Martrat, J., Wesner, S. (eds.): Service Oriented Infrastructures and Cloud Service Platforms for the Enterprise—A Selection of Common Capabilities Validated in Real-Life Business Trials by the BEinGRID Consortium. Springer, Berlin (2010)
21. Erwin, D.W., Snelling, D.F.: Unicore: a grid computing environment. In: Euro-Par '01: Proceedings of the 7th International Euro-Par Conference Manchester on Parallel Processing, pp. 825–834. Springer, London (2001)
22. Evangelinos, C., Hill, C.N.: Cloud computing for parallel scientific HPC applications: feasibility of running coupled atmosphere-ocean climate models on Amazon's EC2. In: Cloud Computing and Its Applications (2008)
23. Foster, I.: What is the grid? A three point checklist (2002). URL http://www.mcs.anl.gov/itf/ Articles/WhatIsTheGrid.pdf (unpublished)
24. Foster, I.: Globus toolkit version 4: Software for service-oriented systems. In: IFIP International Conference on Network and Parallel Computing. LNCS, vol. 3779, pp. 2–13. Springer, Berlin (2005)
25. Foster, I., Geisler, J., Nickless, B., Smith, W., Tuecke, S.: Software infrastructure for the I-WAY high-performance distributed computing experiment. In: Proceedings of the 5th IEEE Symposium on High Performance Distributed Computing, pp. 562–571. Society Press (1996)

26. Foster, I., Kesselman, C.: The Globus project: a status report. Future Gener. Comput. Syst. **15**(56), 607–621 (1999). doi:10.1016/S0167-739X(99)00013-8
27. Foster, I., Kesselman, C., Tuecke, S.: The anatomy of the grid—enabling scalable virtual organizations. Int. J. Supercomput. Appl. **15**, 2001 (2001)
28. Foster, I., Zhao, Y., Raicu, I., Lu, S.: Cloud computing and grid computing 360-degree compared. In: Grid Computing Environments Workshop, GCE '08, pp. 1–10 (2008). doi:10.1109/GCE.2008.4738445
29. Fox, G.C., Pierce, M.: Grids meet too much computing: too much data and never too much simplicity (2007). URL http://grids.ucs.indiana.edu/ptliupages/publications/TooMuchComputingandData.pdf (unpublished)
30. Gentzsch, W.: Sun grid engine: towards creating a compute power grid. In: CCGRID '01: Proceedings of the 1st International Symposium on Cluster Computing and the Grid, p. 35. IEEE Computer Society, Washington (2001)
31. Gottfrid, D.: URL http://open.blogs.nytimes.com/2007/11/01/self-service-prorated-super-computing-fun
32. Gray, J.: Distributed computing economics. Queue **6**(3), 63–68 (2008). doi:10.1145/1394127.1394131
33. Grimshaw, A.S., Wulf, W.A., The Legion Team, C.: The legion vision of a worldwide virtual computer. Commun. ACM **40**(1), 39–45 (1997). doi:10.1145/242857.242867
34. Hadoop: URL http://hadoop.apache.org
35. Hajdu, L., Kocoloski, A., Lauret, J., Miller, M.: Integrating Xgrid into the HENP distributed computing model. J. Phys. Conf. Ser. **119** (2008). doi:10.1088/1742-6596/119/7/072018
36. Hughes, B.: Building computational grids with Apple's Xgrid middleware. In: ACSW Frontiers '06: Proceedings of the 2006 Australasian Workshops on Grid Computing and e-Research, pp. 47–54. Australian Computer Society, Darlinghurst (2006)
37. Kesselman, C., Foster, I.: The Grid: Blueprint for a New Computing Infrastructure. Morgan Kaufmann, San Mateo (1998)
38. Lakshmi, J., Nandy, S.K.: Quality of service for I/O workloads in multicore virtualized servers. In: Cafaro, M., Aloisio, G. (eds.) Grids, Clouds and Virtualization. Springer, Berlin (2010)
39. Laszewski, G.V., Insley, J.A., Foster, I., Bresnahan, J., Kesselman, C., Su, M., Thiebaux, M., Rivers, M.L., Wang, S., Tieman, B., Mcnulty, I.: Real-time analysis, visualization, and steering of microtomography experiments at photon sources. In: Proceedings of the Ninth SIAM Conference on Parallel Processing for Scientific Computing, pp. 22–24. SIAM, Philadelphia (1999)
40. Laure, E., Fisher, S.M., Frohner, A., Grandi, C., Kunszt, P.Z., Krenek, A., Mulmo, O., Pacini, F., Prelz, F., White, J., Barroso, M., Buncic, P., Hemmer, F., Di Meglio, A., Edlund, A.: Programming the Grid using gLite. Tech. Rep. EGEE-PUB-2006-029 (2006)
41. Laure, E., Jones, B.: Enabling grids for e-Science: the EGEE project. Tech. Rep. EGEE-PUB-2009-001. 1 (2008)
42. Li, J., Loo, B.T., Hellerstein, J.M., Kaashoek, M.F., Karger, D.R., Morris, R.: On the feasibility of peer-to-peer web indexing and search. In: IPTPS, pp. 207–215 (2003)
43. Matsuoka, S., Shinjo, S., Aoyagi, M., Sekiguchi, S., Usami, H., Miura, K.: Japanese computational grid research project: Naregi. Proc. IEEE **93**(3), 522–533 (2005)
44. Mersenne Research, I.: URL http://www.mersenne.org
45. Nanda, S., Cker Chiueh, T.: A survey of virtualization technologies. Tech. rep., State University of New York at Stony Brook (2005)
46. Nanda, S., Li, W., Lam, L.C., Chiueh, T.C.: Bird: binary interpretation using runtime disassembly. In: CGO '06: Proceedings of the International Symposium on Code Generation and Optimization, pp. 358–370. IEEE Computer Society, Washington (2006). doi:10.1109/CGO.2006.6
47. Perryman, A.L., Zhang, Q., Soutter, H.H., Rosenfeld, R., McRee, D.E., Olson, A.J., Elder, J.E., Stout, C.D.: Fragment-based screen against HIV protease. Chem. Biol. Drug Des. **75**(3), 257–268 (2010). doi:10.1111/j.1747-0285.2009.00943.x

48. Schick, S.: URL http://blogs.itworldcanada.com/shane/2008/04/22/five-ways-of-defining-cloud-computing/
49. Skype: URL http://www.skype.com
50. Smarr, L., Catlett, C.E.: Metacomputing. Commun. ACM **35**(6), 44–52 (1992). doi:10.1145/129888.129890
51. Smith, W., Foster, I.T., Taylor, V.E.: Predicting application run times using historical information. In: IPPS/SPDP '98: Proceedings of the Workshop on Job Scheduling Strategies for Parallel Processing, pp. 122–142. Springer, London (1998)
52. Thain, D., Tannenbaum, T., Livny, M.: Distributed computing in practice: the condor experience. Concurr. Pract. Exp. **17**(2–4), 323–356 (2005)
53. Wang, L., Zhan, J., Shi, W., Liang, Y., Yuan, L.: In cloud, do MTC or HTC service providers benefit from the economies of scale? In: MTAGS '09: Proceedings of the 2nd Workshop on Many-Task Computing on Grids and Supercomputers, pp. 1–10. ACM, New York (2009). doi:10.1145/1646468.1646475
54. Weiss, A.: Computing in the clouds. Networker **11**(4), 16–25 (2007). doi:10.1145/1327512.1327513
55. Whitaker, A., Shaw, M., Gribble, S.D.: Denali: lightweight virtual machines for distributed and networked applications. In: Proceedings of the USENIX Annual Technical Conference (2002)
56. Xu, M.Q.: Effective metacomputing using lsf multicluster. In: IEEE International Symposium on Cluster Computing and the Grid, p. 100 (2001). doi:10.1109/CCGRID.2001.923181

# Chapter 2
# Quality of Service for I/O Workloads in Multicore Virtualized Servers

J. Lakshmi and S.K. Nandy

**Abstract** Emerging trend of multicore servers promises to be the panacea for all data-center issues with system virtualization as the enabling technology. System virtualization allows one to create virtual replicas of the physical system, over which independent virtual machines can be created, complete with their own, individual operating systems, software, and applications. This provides total system isolation of the virtual machines. Apart from this, the key driver for virtualization adoption in data-centers will be safe virtual machine performance isolation that can be achieved over a consolidated server with shared resources. This chapter identifies the basic requirements for performance isolation of virtual machines on such servers. The consolidation focus is on enterprise workloads that are a mix of compute and I/O intensive workloads. An analysis of prevalent, popular system virtualization technologies is presented with a view toward application performance isolation. Based on the observed lacunae, an end-to-end system virtualization architecture is proposed and evaluated.

## 2.1 Introduction

System virtualization on the emerging multicore servers is a promising technology that has solutions for many of the key data-center issues. Today's data-centers have concerns of curtailing space and power footprint of the computing infrastructure, which the multicore servers favorably address. A typical multicore server has sufficient computing capacity for aggregating several server applications on a single physical machine. The most significant issue with co-hosting multiple-server applications on a single machine is with the software environment of each of the appli-

J. Lakshmi (✉) · S.K. Nandy
SERC, Indian Institute of Science, Bangalore 560012, India
e-mail: jlakshmi@serc.iisc.ernet.in

S.K. Nandy
e-mail: nandy@serc.iisc.ernet.in

M. Cafaro, G. Aloisio (eds.), *Grids, Clouds and Virtualization,*
Computer Communications and Networks,
DOI 10.1007/978-0-85729-049-6_2, © Springer-Verlag London Limited 2011

cations. System virtualization addresses this problem since it enables the creation of virtual replicas of a complete system, over which independent virtual machines (VMs) can be built, complete with their own, individual operating systems, software, and applications. This results in complete software isolation of the VMs, which allows independent applications to be hosted within independent virtual machines on a single physical server.

Apart from the software isolation, the key driver for virtualization adoption in data-centers will be safe virtual machine performance isolation that can be achieved over a consolidated server. This is essential, particularly for enterprise application workloads, like database, mail, and web-based applications that have both CPU and I/O workload components. Current commodity multicore technologies have system virtualization architectures that provide CPU workload isolation. The number of CPU-cores in comparison to I/O interfaces is high in multicore servers. This results in the sharing of I/O devices among independent virtual machines. As a result, this changes the I/O device sharing dynamics when in comparison to dedicated servers, wherein all the resources like the processors, memory, I/O interfaces for disk and network access are architected to be managed by a single OS. On such systems, solutions that optimize or maximize the application usage of the system resources are sufficient to address the performance of the application. When multiple, independent applications are consolidated onto a multicore server, using virtual machines, performance interference caused due to shared resources across multiple VMs adds to the performance challenges. The challenge is in ensuring performance of the independent I/O intensive applications, hosted inside isolated VMs, on the consolidated server while sharing a single I/O device [10].

Prevalent virtualization architectures suffer from the following distinct problems with regard to I/O device virtualization;

1. Device virtualization overheads are high due to which there is a reduction in the total usable bandwidth by an application hosted inside the VM.
2. Prevalent device virtualization architectures are such that sharing of the device causes its access path also to be shared. This causes performance degradation that is dependent on I/O workloads and limits scalability of VMs that can share the I/O device [1].
3. Device access path sharing causes security vulnerabilities for all the VMs sharing the device [35].

These reasons cause variability in application performance that is dependent on the nature of consolidated workloads and the number of VMs sharing the I/O device.

One way to control this variability is to impose necessary Quality of Service (QoS) controls on resource allocation and usage of shared resources. Ideally, the QoS controls should ensure that:

• There is no loss of application performance when hosted on virtualized servers with shared resources.
• Any spare resource is made available to other contending workloads.

The chapter starts with a discussion on the resource specific QoS controls that an application's performance depends on. It then explores the QoS controls for resource allocation and usage in prevalent system virtualization architectures. The focus of this exploration is on the issues of sharing a single NIC across multiple virtual machines (VMs). Based on the observed lacunae, an end-to-end architecture for virtualizing network I/O devices is presented. The proposed architecture is an extension to that of what is recommended in the PCI-SIG IOV specification [21]. The goal of this architecture is to enable fine-grained controls to a VM on the I/O path of a shared device leading to minimization of the loss of usable device bandwidth without loss of application performance. The proposed architecture is designed to allow native access of I/O devices to VMs and provides the device-level QoS controls for managing VM specific device usage. The architecture evaluation is carried out through simulation on a layered queuing network(LQN) [3, 4] model to demonstrate its benefits. The proposed architecture improves application throughput by 60% as in comparison to what is observed on the existing architectures. This performance is closer to the performance observed on nonvirtualized servers. The proposed I/O virtualization architecture meets its design goals and also improves the number of VMs that can share the I/O device. Also, the proposed architecture eliminates some of the shared device associated security vulnerabilities [35].

## 2.2 Application Requirements for Performance Isolation on Shared Resources

Application performance is based on timely availability of the required resources like processors, memory, and I/O devices. The basic guideline for consolidating enterprise servers over multicore virtualized systems is by ensuring availability of required resources as and when required [7]. For the system to be able to do so, the application resource requirements are enumerated using resource requirement (RR) tuples. An RR tuple is an aggregated list of various resources that the application's performance depends on. Thus RR tuple is built using individual resource tuples. Each resource tuple is made up of a list of resource attributes or the attribute tuples. Using this definition, a generic RR tuple can be written as follows:

$$Application(RR) =$$
$$(R1 < A1(Unit, Def, Min, Max), A2(Unit, Def, Min, Max), \dots >,$$
$$R1 < A1(Unit, Def, Min, Max), A2(Unit, Def, Min, Max), \dots >,$$
$$\dots)$$

where:

- *Application(RR)*—Resource requirement tuple of the application.
- *R1*—Name of a resource, viz. processor (CPU), memory, network(NIC), etc.
- *A1*—Name of the attribute of the associated resource. As an example, if A1 represents the CPU speed attribute, it is denoted by the tuple that describes the CPU speed requirements for the application.

- (*Unit, Def, Min, Max*) represent the *Unit* of measurement, *Default*, *Minimum*, and *Maximum* values of the resource attribute.

Using the XML format for resource specification, akin to Globus Resource Specification Language [13], the following example in Fig. 2.1 illustrates the application(RR) for a typical VM that has both compute and I/O workloads.

In the depicted example, the resource tuple for the CPU resource is described by the *<CPU_Resource_Descriptor>* and *</CPU_Resource_Descriptor>* tag pair. Attribute tuples relevant to and associated with this resource are specified using the attribute and value tag pair, within the context of resource tag pair. Each attribute is specified by its *Unit of measurement*, *Default*, *Minimum*, and *Maximum* values that the virtual machine monitor's (VMM's) resource allocator uses for allocating the resource to the VM. In the example, the CPU speed is defined by the attribute tags *<Speed>* and *</Speed>*. The *Unit of Measurement* for CPU speed is mentioned as MHz. The attribute values for *Default*, *Minimum*, and *Maximum* specify the CPU speed required for the desired application performance hosted inside the

```
<Application_RR_descriptor>              <Network_Resource_Descriptor>
<CPU_Resource_Descriptor>               <Speed>
<Speed>                                     <Unit>Mbps</Unit>
    <Unit>MHz</Unit>                        <Default>1000</Default>
    <Default>1800</Default>                 <Minimum>100</Minimum>
    <Minimum>1500</Minimum>                 <Maximum>1000</Maximum>
    <Maximum>2000</Maximum>             </Speed>
</Speed>                                 <Bandwidth>
<NCPU>                                      <Unit>KBps</Unit>
                                           <Default>5000</Default>
    <Default>4</Default>                    <Minimum>5000</Minimum>
    <Minimum>1</Minimum>                    <Maximum>8000</Maximum>
    <Maximum>4</Maximum>               </Bandwidth>
</NCPU>                                      <Unit>KB</Unit>
<L1Cache>                                   <Default>64</Default>
    <Unit>KB</Unit>                         <Minimum>64</Minimum>
    <Default>64</Default>                   <Maximum>64</Maximum>
    <Minimum>64</Minimum>              </Bandwidth>
    <Maximum>64</Maximum>              </Network_Resource_Descriptor>
</L1Cache>                               <Disk_Resource_Descriptor>
</CPU_Resource_Descriptor>               <Size>
<Memory_Resource_Descriptor>                <Unit>MB</Unit>
<Size>                                      <Default>1000</Default>
    <Unit>MB</Unit>                         <Minimum>1000</Minimum>
    <Default>2000</Default>                 <Maximum>1000</Maximum>
    <Minimum>1000</Minimum>            </Size>
    <Maximum>4000</Maximum>            <Bandwidth>
</Size>                                      <Unit>MBps</Unit>
<Bandwidth>                                 <Default>100</Default>
    <Unit>MBps</Unit>                       <Minimum>100</Minimum>
    <Default>6400</Default>                 <Maximum>400</Maximum>
    <Minimum>6400</Minimum>            </Bandwidth>
    <Maximum>6400</Maximum>            </Disk_Resource_Descriptor>
</Bandwidth>                             </Application_RR_Descriptor>
</Memory_Resource_Descriptor>
```

**Fig. 2.1** An example Application Resource Requirement tuple for a VM, expressed in XML

VM. *Default* value specifies the attribute value that the VMM can initially allocate to the VM. On an average, this is the value that the VM is expected to use. The *Minimum* value defines the least value for the attribute that the VM needs to support the guaranteed application performance. The *Maximum* value defines the maximum attribute value that the VM can use while supporting its workload. All the three attribute values can be effectively used if the VMM uses dynamic adaptive resource allocation policies. For each resource, its tuple is specified using attribute value tuples that completely describe the specific resource requirement in terms of the quantity, number of units, and speed of resource access.

On a virtualized server the physical resources of the system are under the control of the VMM. The resource tuples are used by the VMM while allocating or deallocating resources to the VMs. It can be assumed that RR contains values that are derived from the application's performance requirements. In the context of multicore servers, with server consolidation as the goal, each application can be assumed to be hosted in an independent VM which encapsulates the application's environment. Hence, the application's resource tuples can be assumed to be the RR for each VM of the virtualized server. In the case where multiple applications are co-hosted on a single VM, these resource tuples can be arrived at by aggregating the resource requirements of all the applications hosted by the VM.

## 2.3 Prevalent Commodity Virtualization Technologies and QoS Controls for I/O Device Sharing

Commodity virtualization technologies like *Xen* and *Vmware* have made the normal desktop very versatile. A generic architecture of system virtualization, implemented in these systems, is given in Fig. 2.2. The access to CPU resource is native, to all VMs sharing the CPU, for all instructions except the privileged instructions. The privileged instructions are virtualized, i.e., whenever such instructions are executed

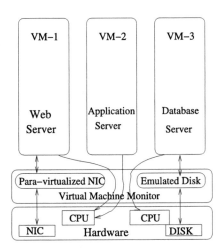

**Fig. 2.2** Generic System Virtualization architecture of prevalent commodity virtualization technologies

from within the VM, they are trapped, and control is passed to the VMM. All I/O instructions fall under the category of privileged instructions. Thus, I/O devices like the Network Interface Card (NIC) and the DISK are treated differently, when virtualized. There are two different, popularly adopted, methods used for virtualizing I/O devices, namely, para-virtualization and emulation [27]. Para-virtualized mode of access is achieved using a virtual device driver along with the physical device driver. A hosting VM or the VMM itself has exclusive, native access to the physical device. Other VMs sharing the device use software-based mechanisms, like the virtual device driver, to access the physical device via the hosting VM or the VMM. In emulated mode of access, each VM sharing the physical device has a device driver that is implemented using emulation over the native device driver hosted by the VMM. Both these modes provide data protection and integrity to independent VMs but suffer from loss of performance and usable device bandwidth. Details of the evaluation are elucidated in the following section. In order to understand the effect of the device virtualization architectures on application performance, experimental results of well-known benchmarks, *httperf* [8] and *netperf* [2], are evaluated. The first experiment is described in Sect. 2.3.1 and explores how virtualization affects application performance. The second experiment, described in Sect. 2.3.2, evaluates the existing QoS constructs in virtualized architectures for their effectiveness in providing application-specific guarantees.

## 2.3.1 Effect of Virtualization on Application Performance

Prevalent commodity virtualization technologies, like *Xen* and *Vmware*, are built over system architectures designed for single OS access. The I/O device architectures of such systems do not support concurrent access to multiple VMs. As a result, the prevailing virtualization architectures support I/O device sharing across multiple VMs using software mechanisms. The result is device sharing along with its access path. Hence, serialization occurs at the device and within the software layers used to access the device.

   In virtualized servers, disk devices are shared differently compared to sharing of NICs. In the case of disk devices, a disk partition is exposed as a filesystem that is exported to a single VM. Any and every operation to this filesystem is from a single VM, and all read and write disk operations are block operations. The data movements to and from the filesystem is synchronized using the filesystem buffer cache that is resident within the VM's address space. The physical data movement is coordinated by the native device drivers within the VMM or the hosted VM, and the para-virtualized or emulated device driver resident in the VM. In the para-virtualized mode, the overheads are due to the movement of data between the device hosting VM and the application VM. In the case of emulation mode of access, the overheads manifest due to the translation of every I/O instruction between the emulated device driver and the native device driver. Due to this nature of I/O activity, VM specific filesystem policies get to be implemented within the software layers of the VMM or the hosting VM. Since the filesystem activity is block based, setting up appropriate block sizes can, to some extent, enable the control of bandwidth and speed requirements on the I/O channel to the disk. However, these controls are still coarse-grained and are insufficient for servers with high consolidation ratios.

For network devices, the existing architecture poses different constraints. Unlike for the disk I/O which is block based, network I/O is packet based, and sharing a single NIC with multiple VMs has intermixed packet streams. This intermixing is transparent to the device and is sorted into per VM stream by the VMM or the hosting VM. Apart from this, every packet is subjected to either instruction translation (emulation) or address translation (para-virtualization) due to virtualization. In both the cases, virtualization techniques build over existing "single-OS over single hardware" model. This degrades application performance.

Throughput studies of standard enterprise benchmarks highlight the effects of virtualization and consolidation based device sharing. Since NIC virtualization puts forth the basic issues with virtualization technologies, an analysis of NIC sharing over application throughput is presented. Figures 2.3a and 2.4a depict the performance of two standard benchmarks, *netperf* [2] and *httperf* [8], wherein the benchmark server is hosted in three different environments, namely nonvirtualized, virtualized, and consolidated servers. The nonvirtualized environment is used to generate the baseline for the metric against which the comparison is made for the performance on virtualized and consolidated server. The virtualized server hosts only one VM wherein the complete environment of the nonvirtualized server is reproduced inside the VM. This environment is used to understand the overheads of virtualization technology. The consolidated server hosts two VMs, similar to the VM of the virtualized server, but with both VMs sharing the same NIC. The consolidated server environment is used to understand the I/O device sharing dynamics on a virtualized server.

For the *netperf* benchmark, *netperf* is the name of the client, and *netserver* is the server component. The study involves execution of the TCP_CRR test of *netperf*. The TCP_CRR test measures the connect-request-response sequence throughput achievable on a server and is similar to the access request used in *http*-based applications. In the case of *httperf* benchmark, the client, called *httperf*, communicates with a standard *http* server using the *http* protocol. In the *httperf* test used, the client allows for specifying the workload in terms of the number of *http* requests to the server in one second, for a given period of time, to generate statistics like the average number of replies received from the server (application throughput), the average response time of a reply, and the network bandwidth utilized for the workload. While *netperf* gives the achievable or achieved throughput, *httperf* gives an average throughput calculated for a subset of samples, executed over a specified period of time, within the given experiment. Hence, *httperf* results give an optimistic estimate which may fall short of expectation in situations where sustained throughput is a requirement.

It is observed from the throughput graphs of *netperf* and *httperf* that there is a significant drop in application throughput as it is moved from nonvirtualized to *Xen* virtualized server. *Xen* virtualization uses para-virtualization mechanism with

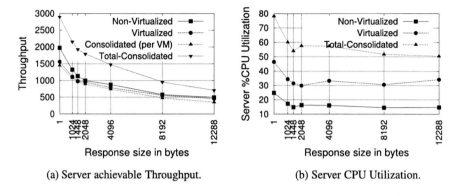

(a) Server achievable Throughput.          (b) Server CPU Utilization.

**Fig. 2.3** *netserver* achievable Throughput and corresponding %CPU Utilization on *Xen* virtualized platform

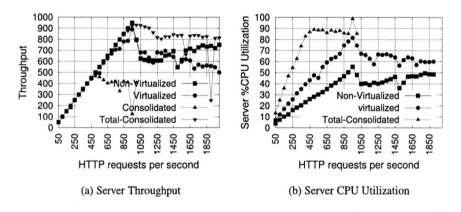

(a) Server Throughput          (b) Server CPU Utilization

**Fig. 2.4** *httperf* server Throughput and %CPU Utilization on *Xen* virtualized single-core server. The hypervisor and the VMs are pinned to the same core

software bridging to virtualize the NIC. The application throughput loss is the overall effect of virtualization overheads. There is a further drop when the application is hosted on a consolidated server with the VMs sharing the NIC. This is obvious, since for the consolidated server, the NIC is now handling twice the amount of traffic in comparison to that of the virtualized server case. It is interesting to note that the virtualization overheads manifest as extra CPU utilization on the virtualized server [17]. This is observed by the CPU utilization graphs of Figs. 2.3b and 2.4b. Both benchmarks indicate increased CPU activity to support the same application throughput. This imposes response latencies leading to application throughput loss and also usable device bandwidth loss for the VM. The noteworthy side effect of this device bandwidth loss, for a VM, is that it is usable by another VM, which is sharing the device. This is noticed in the throughput graphs of the consolidated server for *netperf* benchmark. It is an encouraging fact for consolidating I/O workloads on

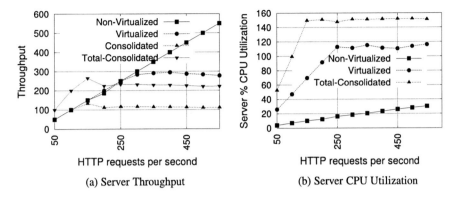

(a) Server Throughput                    (b) Server CPU Utilization

**Fig. 2.5** *httperf* server Throughput and %CPU Utilization on *Vmware-ESXi* virtualized single–core server. The VMs are pinned to a single core, while hypervisor uses all available cores

virtualized servers. However, *httperf* benchmark performance on the consolidated server is not very impressive and suggests further investigation.

Conducting this experiment on *Vmware* virtualization technology produces similar behavior, which is depicted in Fig. 2.5. The *httperf* benchmark tests are conducted on an Intel Core2Duo server with two cores. Unlike the case of *Xen*, pinning of ESXi server (the hypervisor) to a CPU is not allowed. Hence, any CPU utilization measurements for the ESXi hypervisor on *Vmware* show utilizations for all CPUs included. This results in %CPU utilization above 100% in the case of multicore systems. *Vmware-ESXi* server implements NIC virtualization using device emulation. It is observed that the overheads of emulation are comparatively quite high in relation to para-virtualization used in *Xen*. Here also, virtualization of NIC results in using up more CPU to support network traffic on a VM when in comparison to a nonvirtualized server. The other important observation is the loss of application throughput. Device emulation imposes higher service times for packet processing, and hence drastic drop of application throughput is observed in comparison to non-virtualized and para-virtualized systems. In this case 70% drop on the maximum sustained throughput is observed in comparison to the throughput achieved in the nonvirtualized environment. This loss is visible even in the consolidated server case. Interestingly, the total network bandwidth used in the case of consolidated VMs on *Vmware-ESXi* was only 50% of the available bandwidth. Hence, the bottleneck is the CPU resource available to the VMs, since each of the VM was hosted on the same core. It is reasonable to believe that multicores can alleviate the CPU requirement on the consolidated server. On such systems, the CPU requirement of the VMs can be decoupled from that of the VMM by allocating different CPU cores to each of them. Study of *httperf* benchmark on consolidated server with each VM pinned to a different core, for both *Xen* and *Vmware-ESXi* virtualized server, shows otherwise. Application throughput increase is observed in comparison to single-core consolidated server, but this increase still falls short by 10% of what was achieved for the nonvirtualized server. The reason for this shortcoming is because both VMs sharing the NIC also share the access path that is implemented by the Independent

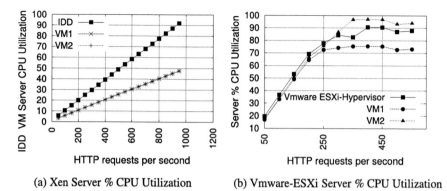

**Fig. 2.6** *httperf* server %CPU Utilization on *Xen* and *Vmware-ESXi* virtualized multicore server. The hypervisor and the VMs are pinned to independent cores

Driver Domain (IDD) in the case of *Xen* and the hypervisor in the case of *Vmware-ESXi* virtualized server. This sharing manifests as serialization and increased CPU utilization of the IDD or the hypervisor, which becomes the bottleneck as the workload increases. Also, this bottleneck restricts the number of VMs that can share the NIC. This is clearly depicted in the graphs of Fig. 2.6. In Fig. 2.6a, it is observed that as the *httperf* workload is increasing, there is a linear increase in the CPU utilization of the VMs as well as the Xen-IDD hosting the NIC. The CPU utilization of the IDD, however, is much more when compared to the CPU utilization of either of the VMs. This is because the IDD is supporting network streams to both the VMs. As a consequence, it is observed that even though there is spare CPU available to the VMs, they cannot support higher throughput since the IDD has exhausted its CPU resource. This indicates that lack of concurrent device with concurrent access imposes serialization constraints on the device and its access path which limits device sharing scalability on virtualized servers. This behavior is also observed in the case of the *Vmware-ESXi* server as is depicted in Fig. 2.6b. However, as in the case of single-core experiments, the CPU Utilization by the hypervisor and the VMs is significantly much higher in comparison to the *Xen* server for the same benchmark workload. This results in poor performance when compared to para-virtualized devices, but yields more unused device bandwidth. As a result, *Vmware-ESXi* server supports higher scalability for sharing the NIC.

The analysis for multicore virtualized server CPU Utilization indicates that even with the availability of required resources, for each of the VMs and the hypervisor, the device sharing architecture has constraints that impose severe restrictions in usable bandwidth and scalability of device sharing. These constraints are specifically due to serialization of device and its access paths. Hence, it is necessary to rearchitect device virtualization to enable concurrent device access to eliminate the bottlenecks evident in device sharing by the VMs.

## 2.3.2 Evaluation of Network QoS Controls

The most noteworthy point of observation of the study in Sect. 2.3.1 is the behavior of each stream of benchmark on the consolidated server. In general, it is observed that there is a further reduction of throughput on the consolidated server in comparison to the single VM on a virtualized server, for both the benchmarks *netperf* and *httperf*, with a marked decrement in the latter case. This indicates the obvious: lack of QoS constraints would lead to severe interference in performance delivered by the device sharing VMs.

The current commodity virtualization technologies like *Xen* and *Vmware* allow for VM specific QoS controls on different resources using different mechanisms. The CPU resource allocations are handled directly by the VMM schedulers like Credit, SEDF, or BVT schedulers of *Xen* [11]. Also, as discussed in [17, 24, 31, 32], the existing CPU resource controls are fine-grained enough to deliver desired performance for CPU-based workloads. The problem is with I/O devices. The access to an I/O device is always through the hypervisor or the driver domain OS kernel to ensure data integrity and protection. The device is never aware as to which VM is using it at any given instance of time; this information and control is managed by the hypervisor or the driver domain. Hence, resource allocation controls with regard to the I/O devices are at a higher abstraction level rather than at the device level, unlike in the case of the CPU resource. These controls are effective for the outgoing streams from the server, since packets that overflow are dropped before reaching the NIC. However, for the incoming stream, the control is ineffective since the decision of accepting or rejecting is made after the packet is received by the NIC. Hence, the controls are coarse-grained and affect the way resource usage is controlled and thereby the application performance. In scenarios where I/O device utilization is pushed to its maximum, limitations of such QoS controls are revealed as loss of usable bandwidth or scalability of sharing, thereby causing unpredictable application performance, as is illustrated in the next section.

To understand the effect of software-based QoS controls for network bandwidth sharing, an experimental analysis of *httperf* benchmark on a consolidated server is presented. The consolidated server hosts two VMs, namely VM1 and VM2, that are sharing a NIC. Each VM hosts one *http* server that responds to a *httperf* client. The *httperf* benchmark is chosen for this study because it allows customization of observation time of the experiment. This is necessary since the bandwidth control mechanisms that are available are based on time-sampled averages and hence, need a certain interval of time to affect application throughput. The experiment involves two studies, one is that of best effort sharing where no QoS is imposed on either of the VMs, and in the second case VM1 is allowed to use the available network bandwidth when VM2 is constrained, by imposing specific QoS value based on the desired application throughput. For both studies, each VM is subjected to equal load from the *httperf* clients.

The performance of consolidated server corresponding to the best effort sharing case is presented in Figs. 2.4a and 2.5a. As it is observed from the graphs, the NIC bandwidth sharing is equal in both the virtualization solutions. When no QoS

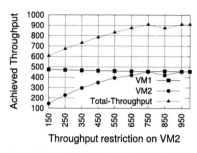 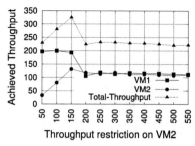

(a) *Xen-IDD Linux* based QoS controls on VM2    (b) *Vmware-ESXi* QoS controls on VM2

**Fig. 2.7** Effect of hypervisor network bandwidth controls on application throughput for consolidated virtualized server hosting two VMs

controls are enforced and each VM has equal demand for the resource, it is shared equally on a best effort basis. In the second study, when bandwidth control is enforced on the VM2, while allowing complete available bandwidth to the VM1, the expected behavior is to see improved throughput for the unconstrained VM1. This is to say, VM1 performance is expected to be better in comparison to the best effort case. Figure 2.7 demonstrates that imposing QoS controls on VM2 does not translate to extra bandwidth availability for the other, unconstrained VM. The reasons for this behavior are a multitude. The most significant ones being the virtualization overhead in terms of the CPU resource required by the VMM or the hosting VM to support I/O workload, serialization of the resource and its access path, lack of control on the device for the VM specific incoming network stream, and lastly, higher priority to the incoming stream over the outgoing stream at the device. All these lead to unpredictable application performance inspite of applying appropriate QoS controls. Also, it is interesting to note that the variation in performance is dependent on the nature of the consolidated workloads. This performance variation affects all the consolidated workloads and makes the application guarantee weak. On multi-core servers hosting many consolidated workloads of a datacenter, indeterminate performance is definitely not acceptable. Also, since virtual device is an abstraction supported in software, device usage controls are coarse grained and hence ineffective. This could lead to an easy denial of service attack on a consolidated server with shared devices.

The bandwidth controls enforced are based on the following principle. For each of the virtualization technologies used, i.e., *Xen* and *Vmware*, the network bandwidth used by a single VM to support different *httperf* request rates, without performance loss, is measured. These bandwidth measurements are used to apply control on the outgoing traffic from VM2. Currently, the available controls allow constraints only on the outgoing traffic. On the incoming traffic, ideally the control should be applied at the device so that any packet causing overflow is dropped before reception. Such controls are not available at present. Instead, in *Xen*, at least one can use the *netfilter module's* stream-based controls after receiving the packet. This does not serve the purpose, because by receiving a packet that could potentially be dropped later, the device bandwidth is anyway wasted. Hence, the study involves using only the outgoing traffic controls for the constrained VM.

The selection of different range of workloads, for each of the virtualized server, is based on the maximum throughput that each can support in a consolidated server environment. For each QoS control, the maximum throughput achieved, without loss, by each of the VM, is plotted in the graphs of Figs. 2.7a and b. In these figures, the $x$-axis represents the *httperf* request rate based on which the network bandwidth control was applied on the VM2, and the $y$-axis represents the application throughput achieved by each of the VMs. In the case of *Xen*, *Linux tc* utility of the *netfilter* module [34] is used to establish appropriate bandwidth controls. Specifically, each traffic stream from the VMs is defined using *htb* class with *tbf* queue discipline with the desired bandwidth control. Each queue is configured with a burst value to support a maximum of 10 extra packets. In the case of *Vmware-ESXi* server, the *Veam Monitor* controls for network bandwidth are used and populated with the same QoS controls as is done for the *Xen* server.

Based on the behavior of the benchmarks, following bottlenecks are identified for sharing network I/O device across multiple VMs on *Xen* or *Vmware-ESXi* virtualized server.

- Virtualization increases the device utilization overheads, which leads to increased CPU utilization of the hypervisor or the IDD hosting the device.
- Virtualization overheads cause loss of device bandwidth utilization from inside a VM. Consolidation improves the overall device bandwidth utilization but further adds to CPU utilization of the VMM and IDD. Also, if the VMM and IDD do not support concurrent device access APIs, they themselves become the bottlenecks for sharing the device.
- QoS features for regulating incoming and outgoing traffic are currently implemented in the software stack. Uncontrolled incoming traffic at the device, to a VM that is sharing a network device, can severely impact the performance of other VMs because the decision to drop an incoming packet is taken after the device has received the packet. This could potentially cause a denial of service attack on the VMs sharing the device.

Based on the above study, a device virtualization architecture is proposed and described in the following sections. The proposal is an extension to I/O virtualization architecture, beyond what is recommended by the PCI-SIG IOV specification [21]. The PCI-SIG IOV specification defines the rudiments for making I/O devices virtualization aware. On the multicore servers with server consolidation as the goal, particularly in the enterprise segment, being able to support multiple virtual I/O devices on a single physical device is a necessity. High-speed network devices, like 10-Gbps NICs, are available in the market. Pushing such devices to even 80% utilization needs fine-grained resource management at the device level. The basic goal of the proposed architecture is to be able to support finer levels of QoS controls, without compromising on the device utilization. The architecture is designed to enable native access of I/O devices to the VMs and provide device-level QoS hooks for controlling VM specific device usage. The architecture aims to reduce network

I/O device access latency and enable improvement in effective usable bandwidth in virtualized systems by addressing the following issues:

- Separating device management issues from device access issues.
- Allowing native access of a device to a VM by supporting concurrent device access and eliminating hypervisor/IDD from the path of device access.
- Enable fine-grained resource usage controls at the device.

In the remaining part of the chapter, we bring out the need for extending I/O device virtualization architecture in Sect. 2.4. Section 2.5 highlights the issues in sharing of the I/O device and its access path in prevalent virtualization architectures leading to a detailed description of the proposed architecture to overcome the bottlenecks. *Xen* virtualization architecture is taken as the reference model for the proposed architecture. In the subsequent part of the section, a complete description of the network packet work-flow for the proposed architecture is presented. These work-flows form a basis for generating the LQN model that is used in the simulation studies for architecture evaluation described in Sect. 2.6. A brief description of the LQN model generation and detailed presentation of simulation results is covered in Sect. 2.7. Finally, in Sect. 2.8 the chapter conclusion highlights on the benefits of the architecture.

## 2.4 Review of I/O Virtualization Techniques

Virtualization technologies encompass a variety of mechanisms to decouple the system architecture and the user-perceived behavior of hardware and software resources. Among the prevalent technologies, there are two basic modes of virtualization, namely, full system virtualization as in *Vmware* [15] and para-virtualization as in *Xen* [11]. In full system virtualization complete hardware is replicated virtually. Instruction emulation is used to support multiple architectures. The advantage of full system virtualization is that it enables unmodified Guest operating systems (GuestOS) to execute on the VM. Since it adopts instruction emulation, it tends to have high performance overheads as observed in the experimental studies described earlier. In Para-virtualization the GuestOS is also modified suitably to run concurrently with other VMs on the same hardware. Hence, it is more efficient and offers lower performance overheads. In either case, system virtualization is enabled by a layer called the virtual machine monitor (VMM), also known as the hypervisor, that provides the resource management functionality across multiple VMs. I/O virtualization started with dedicated I/O devices assigned to a VM and has now evolved to device sharing across multiple VMs through virtualized software interfaces [27]. A dedicated software entity, called the I/O domain, is used to perform physical device management [9, 12]. The I/O domain can be part of the VMM or be an independent domain, like the independent driver domain (IDD) of *Xen*. In the case of IDD, the I/O devices are private to the domain, and memory accesses by the devices are restricted to the IDD. Any application in a VM seeking access to the device has to route the request through the IDD, and the request has to pass through the address translation barriers of the IDD and VM [14, 19, 20, 22].

Recent publications on concurrent direct network access (CDNA) [23] and scalable self-virtualizing network interface [16] are similar to the proposed work in the sense that they explore the I/O virtualization issues on the multicore platforms and provision for concurrent device access. However, the scalable self-virtualizing interface describes assigning a specific core for network I/O processing on the virtual interface and exploits multiple cores on embedded network processors for this. The authors do not detail how the address translation issues are handled, particularly in the case of virtualized environments. CDNA is architecturally closer to our architecture since it addresses concurrent device access by multiple VMs. CDNA relies on per VM Receive (Rx) and Transmit (Tx) ring buffers to manage VM specific network data. The VMM handles the virtual interrupts, and the *Xen* implementation still uses IDD to share the I/O device. Also, authors do not address the performance interference due to uncontrolled data reception by the device nor do they discuss the need for addressing the QoS controls at the device level.

The proposed architecture addresses these and suggests pushing the basic constructs to assign QoS attributes like required bandwidth and priority into the device to get fine-grained control on interference effects. Also, the proposed architecture has it basis in *exokernel's* [6] philosophy of separating device management from protection. In *exokernel*, the idea was to extend native device access to applications with the *exokernel* providing the protection. In the proposed approach, the extension of native device access is with the VM, the protection being managed by the VMM and the device collectively. A VM is assumed to be running the traditional GuestOS without any modifications with native device drivers. This is a strong point in support of legacy environments without any need for code modification. Further, the PCI-SIG community has realized the need for I/O device virtualization and has come out with the IOV specification to deal with it. The IOV specification, however, talks about device features to allow native access to virtual device interfaces, through the use of I/O page tables, virtual device identifiers, and virtual device-specific interrupts. The specification presumes that QoS is a software feature and does not address this. Many implementations adhering to the IOV specification are now being introduced in the market by Intel [18], Neterion [25], NetXen [26], Solarflare [33], etc. Apart from these, the Crossbow [28] suite from SUN Microsystems talks about this kind of resource provisioning. However, Crossbow is a software stack over a standard IOV complaint hardware. The results published using any of these products are exciting in terms of the performance achieved. These devices when used within the prevalent virtualization technologies need to still address the issue of provisioning QoS controls on the device. Lack of such controls, as illustrated by the previously described experimental studies, cause performance degradation and interference that is dependent on the workloads sharing the device.

## 2.5 Enhancement to I/O Virtualization Architecture

The analysis of prevalent commodity virtualization technologies in Sect. 2.3 clearly highlights the issues that need to be addressed while sharing I/O devices across independent VMs on multicore virtualized servers. It is also observed that while para-virtualization offers better performance for the application, emulation is an alternative for improved consolidation. The goals are seemingly orthogonal since current technologies build over virtualization unaware I/O devices. The proposed architecture takes a consolidated perspective of merging these two goals, that of ensuring application performance without losing out on the device utilization by taking advantage of virtualization aware I/O devices and rearchitecting the end-to-end virtualization architecture to deliver the benefits. In order to understand the benefits of the proposed architecture, the *Xen*-based para-virtualization architecture for I/O devices is taken as the reference model. In the existing *Xen* virtualization architecture, analysis of the network packet work-flow highlights following bottlenecks:

- Since the NIC device is shared, the device memory behaves like a common memory for all the contending VMs accessing the device. One misbehaving VM can ensure deprivation leading to data loss for another VM.
- The *Xen-IDD* is the critical section for all the VMs sharing the device. IDD incurs processing overheads for every network operation executed on behalf of each VM. Current IDD implementations do not have any hooks for controlling the overheads on per VM basis. Lack of such controls leads to performance interference in the device sharing VMs.
- Every network packet has to cross the address translation barrier of VMM to IDD to VM and vice-versa. This happens because of lack of separation of device management issues from device access issues. Service overheads of this stage-wise data movement cause drop in effective utilized device bandwidth. In multicore servers with scarce I/O devices, this would mean having high-bandwidth underutilized devices and low-throughput applications on the consolidated server.

To overcome the above-listed drawbacks, the proposed architecture enhances I/O device virtualization to enable separation of device management from device access. This is done by building device protection mechanisms into the physical device and managed by the VMM. As an example, for the case of NIC, the VMM recognizes the destination VM of an incoming packet by the interrupt raised by the device and forwards it to the appropriate VM. The VM then processes the packet as it would do so in the case of nonvirtualized environment. Thus, device access and scheduling of device communication are managed by the VM that is using it. The identity for access is managed by the VMM. This eliminates the intermediary VMM/IDD on the device access path and reduces I/O service time, which improves the application performance on virtualized servers and also the usable device bandwidth which results in improved consolidation. In the following subsections we describe the NIC I/O virtualization architecture, keeping the above goals in mind, and suggest how the system software layers of the VMM and the GuestOS inside the VM should use the NIC hardware that is enabled for QoS-based concurrent device access.

### 2.5.1 Proposed I/O Virtualization Architecture Description

Figure 2.8 gives a block schematic of the proposed I/O virtualization architecture. The picture depicts a NIC card that can be housed within a multicore server. The card has a controller that manages the DMA transfer to and from the device memory. The standard device memory is now replaced by a partitionable memory supported with $n$ sets of device registers. A set of $m$ memory partitions, where $m \leq n$, along with device registers, forms the virtual-NICs (vNICs). The device memory is reconfigurable, i.e., dynamically partitionable, and the VM's QoS requirements drive the sizing of the memory partition of a vNIC. The advantage of having a dynamically partitionable device memory is that any unused memory can be easily extended into or reduced from a vNIC in order to support adaptive QoS specifications. The NIC identifies the destination VM of an arriving packet, based on the logical device address associated with it. A simple implementation is to allow a single physical NIC to support multiple MAC address associations. Each MAC address then represents a vNIC, and a vNIC request is identified by generating a message-signaled interrupt (MSI). The number of MAC addresses and interrupts supported by the controller restricts the number of vNICs that can be exported. Although the finite number of physical resources on the NIC restricts the number of vNICs that can be exported, judicious use of native and para-virtualized access to the vNICs, based on the QoS guarantees a VM needs to honor, can overcome the limitation. A VM that has to support stringent QoS guarantees can choose to use native access to the vNIC, whereas those VMs that are looking for best-effort NIC access can be allowed para-virtualized access to a vNIC. The VMM can aid in setting up the appropriate hosting connections based on the requested QoS requirements. The architecture can be realized with the following enhancements:

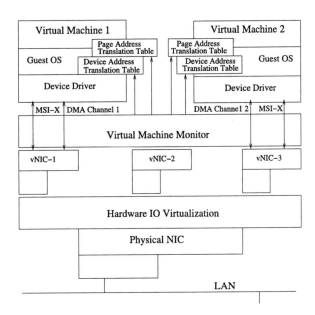

**Fig. 2.8** NIC architecture supporting independent reconfigurable virtual-NICs

**Virtual-NIC**

In order to define vNIC, the physical device should support timesharing in hardware. For a NIC, this can be achieved by using MSI and dynamically partitionable device memory. These form the basic constructs to define a virtual device on a physical device as depicted in Fig. 2.8. Each virtual device has a specific logical device address, like the MAC address in case of NICs, based on which the MSI is routed. Dedicated DMA channels, a specific set of device registers, and a partition of the device memory are part of the virtual device interface which is exported to a VM when it is started. This virtual interface is called the vNIC which forms a restricted address space on the device for the VM to use and remains in possession of the VM until it is active or relinquishes the device. The VMM sets up the device page translation table, mapping the physical device address of the vNIC into the virtual memory of the importing VM, during the vNIC creation and initialization. The device page translation table is given read-only access to the VM and hence forms a significant security provisioning on the device. This prohibits any corrupt device driver of the VM GuestOs to affect other VMs sharing the device or the VMM itself. Also, for high-speed NIC devices, the partitionable memory of the vNIC is useful in setting up large receive and segment offload capabilities specific to each vNIC and thus customizes the sizing of each vNIC based on the QoS requirements of the VM.

**Accessing Virtual-NIC**

To access the vNIC, the native device driver hosted inside the VM replaces the IDD layer. This device driver manipulates the restricted device address space which is exported through the vNIC interface by the VMM. The VMM identifies and forwards the device interrupt to the destination VM. The GuestOS of the VM handles the I/O access and thus directly accounts for the resource usage it incurs. This eliminates the performance interference when the IDD handles multiple VM requests to a shared device. Also, direct access of vNIC to the VM reduces the service time on the I/O accesses. This results in better bandwidth utilization. With the vNIC interface, data transfer is handled by the VM. The VM sets up the Rx/Tx descriptor rings within its address space and makes a request to the VMM for initializing the I/O page translation table during bootup. The device driver uses this table along with the device address translation table and does DMA directly into the VM's address space.

**QoS and Virtual-NIC**

The device memory partition acts as a dedicated device buffer for each of the VMs. With appropriate logic on the NIC card, QoS-specific service level agreements (SLAs) can be easily implemented on the device that translates to bandwidth restrictions and VM-based processing priority. The key is being able to identify the incoming packet to the corresponding VM. This is done by the NIC based on the packet's associated logical device address. The NIC controller decides on whether to accept or reject the incoming packet based on the bandwidth specification or the current free memory available with the destination vNIC of the packet. This gives a fine-grained control on the incoming traffic and helps reduce the interference effects. The outbound traffic can be controlled by the VM itself, as is done in the existing architectures.

**Security and Virtual-NIC**

Each vNIC is carved out as a device partition, based on the device requirement specification of the VM. By using appropriate microarchitecture and hardware constructs it can be ensured that a VM does not monopolize device usage and cause denial of service attack to other VMs sharing the device. The architecture allows for unmodified GuestOS on a VM. Hence the security is verified and built outside the VM, i.e., within the VMM. Allowing native device driver within the VM for the vNIC not only enhances the performance but also allows for easy trapping of the device driver errors by the VMM. This enables for building robust recovery mechanisms for the VM. The model also eliminates sharing of the device access path by allowing direct access to the vNIC by the VM and thereby eliminates the associated failures [35].

With these constructs, the virtualized NIC is now enabled for carving out secure, customized vNICs for each VM, based on its QoS requirements, and supports native device access to the GuestOS of the VM.

## 2.5.2 Network Packet Work-Flow Using the Virtualized I/O Architecture

With the proposed I/O device virtualization architecture, each VM gets safe, direct access to the shared I/O device without having to route the request through the IDD. Only the device interrupts are routed through the VMM. In Figs. 2.9a and b, the workflow for network data reception and transmission using the described device virtualization architecture is depicted. When a packet arrives at the NIC, it deciphers the destination address of the packet, checks if it is a valid destination, then copies the packet into the vNIC's portion of the device memory and issues DMA request to the destination VM based on the vNIC's priority. On completion of the DMA request, the device raises an interrupt. The VMM intercepts the interrupt, determines the destination VM, and forwards the interrupt to the VM. The VM's device driver then receives the data from the VM specific device descriptor rings as it would do in the case of nonvirtualized server. In the case of transmission, the device driver that is resident in GuestOS of the VM does a DMA transfer of the data directly into the vNIC's mapped memory and sets the appropriate registers to

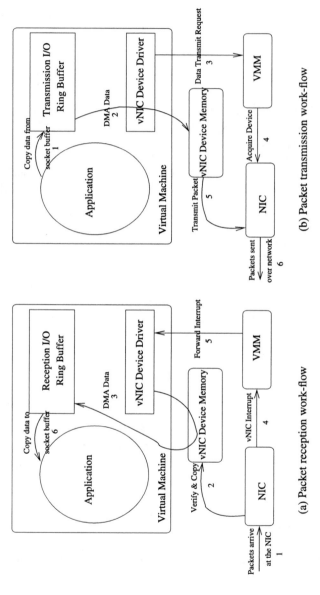

**Fig. 2.9** Workflow of network I/O communication with improvised I/O device virtualization architecture

initiate data transmission. The NIC transmits this data based on the vNICs properties like speed, bandwidth, and priority. It may be worth noting here that the code
changes to support this architecture in the existing implementation will be minimal.
Each VM can use the native device driver for its vNIC. This device driver is the
standard device driver for the IOV complaint devices with the only difference that
it can only access restricted device address. The device access restrictions in terms
of memory, DMA channels, interrupt line, and device register sets are setup by the
VMM when the VM requests for a virtual device. With the virtual device interface,
the VMM now only has to implement the virtual device interrupts.

## 2.6  Evaluation of Proposed Architecture

Since the architecture involves the design of a new NIC and a change in both VMM
and the device handling code inside the VM's GuestOS, evaluation of the architecture is carried out using simulation based on LQN model of the architecture. In
LQN models, functional components of the architecture workflow are represented
as server entries. Service of each entry is rendered on a resource. End-to-end workflow is enacted using entry interactions. The LQN models capture the contention at
the resource or software component using service queues. The reason for choosing
LQN-based modeling is twofold. First, there is a lack of appropriate system simulation tools that allow incorporating design of new hardware along with VMM and
GuestOS changes. Second, LQN models are intuitive queuing models that enable
capturing of the device and software contention and associated serialization in the
end-to-end workflow, right from the application to the device including the intermediate layers of the VM, IDD, and VMM. With appropriate profiling tools, the
LQN models are fairly easy to build and are effective in capturing the causes of
bottlenecks in the access path. For complete details on general description of LQN
modeling and simulation, the reader may refer to [3–5].

### 2.6.1  LQN Model for the Proposed Architecture

LQN models can be generated based on the network packet receive and transmit
workflows, manually, using the LQNDEF [3] software developed at the RADS laboratory of *Carleton University*. In the chapter, results generated for the LQN model
corresponding to the *httperf* benchmark are presented for analysis, since the bottleneck issues are prominent for this benchmark. For complete details on the generation
of the LQN models for the *httperf* benchmark and validation of the models against
experimental data, readers may refer to [29, 30]. Three assumptions are made while
generating the LQN models used for this analysis, namely:

- The service times established at each of the entries constituting the LQN model
  are populated based on the service times measured for an *http* request, instead of

a *TCP* packet. While it is feasible to model packet level contention, the reason for choosing request level contention was to enable measurement of the model throughput in terms of the number of satisfied requests. The model validation results demonstrate that there is no significant loss or gain (<1%) of throughput because of this.

- The experimental results for *httperf* benchmark illustrated in Sect. 2.3 are carried out with varying request rates for a single specified file. In this mode of execution, the file that is fetched as a reply to each of the *http* request, remains constant. Hence the measured service time to process each request remains constant. Also, for the chosen mode of execution of the *httperf* benchmark, the arrival request rate is observed to be uniform. Hence, the service times and arrival rates populated on the LQN model are modeled as deterministic.
- The service time for all device activities that are assumed to be executed in hardware, in the proposed architecture and modeled as separate entities in the LQN model, is set to be significantly low ($10^{-10}$ seconds). For the rest of the software entries, the service times are derived based on the measurements made for the nonvirtualized servers. This is justified since the proposed architecture gives native access to the device from within the VM which is assumed to be running the same GuestOS as is used for the nonvirtualized server.

In general it is observed that the maximum throughput observed using the LQN model is higher than the experimental observations. The reason for this is simple. For every packet received or transmitted in *Linux*, there are several layers of the network stack that each packet has to pass through. The time taken to traverse this passage is recorded by the profiler as the service time. In the real system, to match the difference between the device speed and CPU speed, appropriate memory buffers (TCP transmit and receive buffers of *Linux* kernel) are maintained. The sizing of these buffers affects the observed application throughput. Observed throughputs are higher for larger buffer sizes. This trend is maintained to the point until the device can handle the rate of network traffic. Once device saturation occurs, the failure behavior usually results in a sudden drop in application throughput. While setting up the LQN model, the maximum permissible default buffer size was used in the simulator (which is more than three times than what was set on the experimental system). This is normally the adopted practice since in throughput studies the interest is to understand the limits of the model for those service times that make the contention predominant. This gives an idea on the upper bound of application throughput on a system with maximum possible resources for the service times possible within the desired architecture. The basic idea is to eliminate buffer size constraint in the simulation environment. While it is true that for the proposed architecture in which native access to the I/O device is provided, the maximum throughput that can be achieved, in reality, cannot exceed that of the maximum throughput achieved in the case of nonvirtualized server, the results observed using simulations are contradictory. This is because in the simulation environment, the buffer sizes used were much larger than the experimental system. Hence, to make the comparison fair, normalization of simulation results for existing architecture is carried out. To normalize, the LQN model of existing *Xen* architecture is built, and simulation results are generated.

These results are verified and validated for correctness with that of observed experimental results. After this, all comparisons for the proposed architecture are made using the simulation results of the existing architecture rather than the experimental results.

## 2.7 Simulation and Results

The proposed architecture is evaluated using the *parasrvn* simulator of the *LQNS* software package [3].The architecture is evaluated for multicore virtualized servers since the illustrated device sharing dynamics are expected to be pertinent to such systems. The LQN model built for this study consists of one VMM and two VMs, and each is pinned to an independent core. In order to compare the performance of the proposed end-to-end architecture within the simulation environment, validation of the LQN model for the existing *Xen* architecture for a multicore server is carried out. Figure 2.10 depicts the results of achievable throughput and server CPU utilization for a multicore *Xen* server with two VMs consolidated. The throughput graph for both the VMs is similar and appears overlapped in the chart. As it can be noted from Figs. 2.10a and b, in a multicore environment with *Xen-IDD*, VM1 and VM2 each pinned to a core, and each VM servicing one *httperf* stream, the maximum throughput, without loss, achievable per stream is 950 requests/s as against 450 requests/s in the case of single core. But, for the maximum throughput, it is observed that the *Xen-IDD*, which is hosting the NIC of the server, the CPU utilization saturates. This indicates that further increase in application throughput is impossible since the processor core serving the *Xen-IDD* has no computing power left. Figure 2.11 shows these statistics for a similar situation but with the proposed I/O virtualization architecture. As one can observe from Fig. 2.11a, the maximum throughput achievable now per VM increases to 1500 requests/s. This is an increase of application throughput by about 60%. The total throughput achievable at the NIC, derived from consolidating the throughput of both the VMs, also increases by 60% in comparison to what was achieved on the existing *Xen* architecture.

Also, from Fig. 2.11b it is observed that the CPU utilization of the *IDD* or the hypervisor has considerably reduced and remains bounded by an upper limit. The reason for this behavior is that the NIC is now handling the identity of the packet destination. Also, in the existing model, bridging software, which routes the packets to a VM and has a substantial overhead, is eliminated in the proposed architecture. The effect is a reduction in the processing time that the *IDD* spends on behalf of each VM. It is also noticed that since the VMM is now spending almost constant time on I/O requests on behalf of the VMs, there is an elimination of performance interference due to varying workloads. This improves the scalability of sharing the device across VMs. With the proposed architecture, each VM is now accountable for all the resource consumption, thereby leading to better QoS controls.

The next evaluation of the proposed architecture is for QoS controls on the network bandwidth. Since the architecture is implemented using LQN model, certain

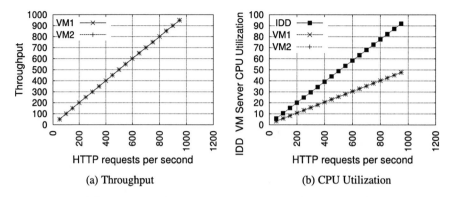

**Fig. 2.10** Maximum throughput achievable per *httperf* stream and CPU utilization for existing *Xen* architecture on a multicore server hosting two VMs, each servicing one of the *httperf* stream. The IDD, VM1, and VM2 are pinned to independent cores

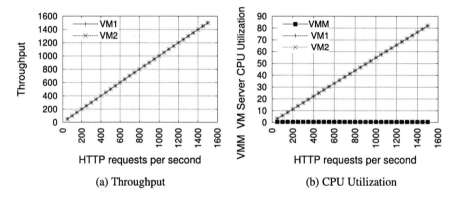

**Fig. 2.11** Maximum achievable throughput and CPU utilization charts for a multicore virtualized server incorporating the proposed I/O virtualization architecture and hosting two VMs, pinned to different cores, each servicing one *httperf* stream

modeling assumptions are made to simulate the network bandwidth controls as implemented in the *netfilter* module of *Linux*. LQN model is basically a queuing model wherein any node (also called entry in *parasrvn* notation) of the queue is described using three parameters, namely, the arrival rate, the service time, and the think time. The arrival rate models the rate of input requests at the entry, service time represents the time the entry takes to process the request before forwarding to the next entry or replying back to the requesting entry, and think time denotes the time before which the entry actually services the request. The think time parameter is useful to model policies like bandwidth restrictions, time-sharing intervals, periodic processing, etc. The LQN model is basically a directed acyclic graph that captures the complete workflow. Hence, the arrival rate is set for the source entry and in this case represents the rate of request arrival at the network interface of the virtualized server. The service time represents the resource time used for servicing the request by the entry of LQN model, and think time is used to model bandwidth restriction. For example, to model 250 requests/second bandwidth restriction, the think time derived is 1/250 seconds. This ensures that the entry will only process 250 requests/second and anything extra will be queued or dropped. The next parameter to model is the burst parameter of the bandwidth control mechanism in *Linux netfilter* module. In *Linux netfilter* module, once the bandwidth limit is reached, packet loss occurs. The bandwidth control mechanism also has a burst parameter that allows for some extra packet delivery on the channel, over and above that of imposed bandwidth restriction. By setting the burst rate sufficiently low, equivalent to 10 packets, which is also the minimum that is permissible, it is ensured that the bandwidth control on the constrained channel is tight. The *HTML* page that is requested in the experiments requires fourteen packets to complete a successful request. Since there is no feature in LQN model to associate the burst parameter of *netfilter*, the QoS experiments were carried out by setting the burst rate to 10 packets. This ensures that for the request that exceeds the configured bandwidth, control fails, and the throughput reported takes into account the desired behavior. Thus, think time setting in LQN model is more restrictive than the *netfilter*. However, since the think time value is based on the deterministic request rate parameter that defines the bandwidth constraint, it still produces equivalent results, and this has been validated against observed experimental values [29].

The following graphs in Fig. 2.12 depict the effect of not imposing (Fig. 2.12a) and imposing network bandwidth QoS controls on the incoming stream of VM2 (Fig. 2.12b), in the proposed architecture. The simulations are conducted on a single core server to keep the achievable throughput range within reasonable simulation time. As it can be observed from the graphs of Fig. 2.12a, for the best effort service, the maximum throughput, without loss, achieved by either of the VMs on the consolidated server is equal, indicating a fair share of the resource. The graphs of the Fig. 2.12b show that, unlike as in the case of existing architectures, the QoS constraints, when moved to device level, allow the usage of available bandwidth by the unconstrained channel. In the figure, VM2 is constrained to allow requests starting from 150 requests/second to 950 requests/second, and VM1 is unconstrained. Since

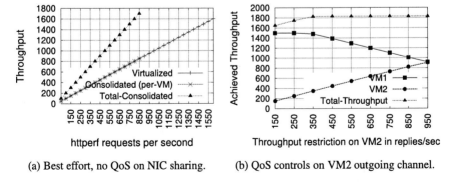

(a) Best effort, no QoS on NIC sharing.        (b) QoS controls on VM2 outgoing channel.

**Fig. 2.12** Throughput achieved before and after imposing QoS controls on VM2 of the proposed architecture

the NIC is discarding requests to VM2 that are above the specified request rate, VM1 can use the available bandwidth, and hence higher throughput (1500 replies/sec) on VM1 is achievable. As the bandwidth control on VM2 is relaxed, it is noticed that the throughput graphs start converging toward each other and finally merge to that of the best effort case. The bandwidth control on the incoming stream also works to our advantage on the *http* traffic, because by discarding the request at the device itself, the server and hence the associated resources are spared to respond on requests that will eventually be dropped because of bandwidth controls. This control on the device also acts as a strong deterrent for any denial of service type of attacks. The other observation is that when multiple VMs are sharing the NIC, the maximum bandwidth achievable on the unconstrained channel is less (<10%) than that which is achieved by the isolated VM. Further reduction on this loss is possible by applying channel-based priority and bandwidth control on the outgoing channel of the constrained VM. The outgoing channel constraints are easily achievable by using existing mechanisms such as those available in the *netfilter* module of *Linux* [34]. The important point to note here is that with faster and higher-bandwidth NIC devices, judicious use of large receive and segment offload buffers can lead to higher device utilization without compromising the VM's performance.

## 2.8 Conclusion

In this chapter, we described how the lack of virtualization awareness in I/O devices can lead to latency overheads on the I/O path and also cause security vulnerabilities. In addition to this, the intermixing of device management and data protection issues further increases the latency. This results in reducing the effective usable bandwidth of the device. Also, lack of appropriate device-sharing control mechanisms, at the device level, leads to loss in bandwidth, causes performance interference on the device sharing VMs, and makes the virtualization software the most vulnerable component of the consolidated server. To address these issues, I/O device virtualization architecture is proposed. The architecture is an extension to the PCI-SIG IOV specification. The architecture evaluation is done by capturing it as an LQN model and analyzing using simulation of the model. The simulation results show a utilization benefit of about 60%, without enforcing any QoS guarantees or performing any software optimization on the I/O path. The proposed architecture also improves the security and scalability of VMs sharing the NIC. It is demonstrated that by moving the QoS controls to the shared device, the unused bandwidth is made available to the unconstrained VM, unlike the case in prevalent technologies. Although the evaluation is done for para-virtualized systems like *Xen*, it is reasonable to expect that the ideas presented would benefit fully virtualized systems like *Vmware* since the architecture enables elimination of the common software entity by providing native device access to the GuestOS of the VM.

**Acknowledgements** Credits for this work are due to all those unknown reviewers who have meticulously pointed out deficiencies and improvements over several rounds of reviews and also to the summer interns who have enthusiastically carried out the numerous experimental work that helped validate the simulation results.

# Appendix

Layered Queuing Network (LQN) models are the queuing models designed to capture the interdependencies in layered systems. The complete system is described by a set of operations carried out over a set of resources. Every operation requires one or more resources for execution. The LQN model defines an architectural and resource context for each operation. The architectural context defines the initiating event for the operation (execution trigger), when the execution should begin (execution timing) and when it should complete (completion trigger). Based on the semantics of the architectural context, the operation uses resources to carry out its activities, which is defined by its resource context. A resource can be a software entity or a hardware unit involved in actual execution of the operation. Each resource is associated with a queue with a discipline that enforces the order of resource use by the tasks. In layered systems, execution of an activity is carried out by a structured order of operations over resources organized in different layers. An LQN model is necessarily an acyclic graph of all possible sequences of requests to avoid the issue of resource deadlocks. LQNs are very intuitive in capturing resource contentions and thereby the performance implications on a layered system. These models are quite common in practice for modeling software system performance.

The LQN models used in this chapter to evaluate I/O virtualization architecture for the *httperf* benchmark are generated using the software developed at the RADS Laboratory of Carleton University. Complete details of the software, tools, and the associated documentation can be found on their website [3].

A short description of the LQN models generated for the proposed I/O virtualization architecture and *Xen* is provided here. The I/O virtualization issues are prominent for the *httperf* benchmark, and hence LQN models that capture the end-to-end architecture are generated for analyzing the issues. The diagrams in Figs. 2.13 and 2.14 depict the LQN models generated for a consolidated Xen server and the proposed I/O virtualization architecture, hosting two VMs. The model has two *httperf* streams accessing *http* servers hosted on different VMs. The model captures the scenario for a multicore system. In these models, each rectangular box represents the conceptual functional entity that is active in the receive or the transmit path of the network packet workflows depicted in Fig. 2.9, to complete one *httperf* request–reply sequence. To make the LQN model simpler, a few assumptions are made:

1. While in reality every *http* request is broken into a sequence of packets that are passed through various layers of OS, on an LQN model it is captured as a single service request. This allows for throughput measurements on the model in terms of satisfied *http* requests. This is the unit of measurement for the *httperf* benchmark. By aggregating contention issues from packet level to request level, the throughput measurements tend to be optimistic than what is observed in actual experiments.

2. The service time associated with the transmit/receive operation is consolidated to represent the sending of all the packets composing the *http* request. Because of this assumption, the results of the simulation tend to give upper bounds on

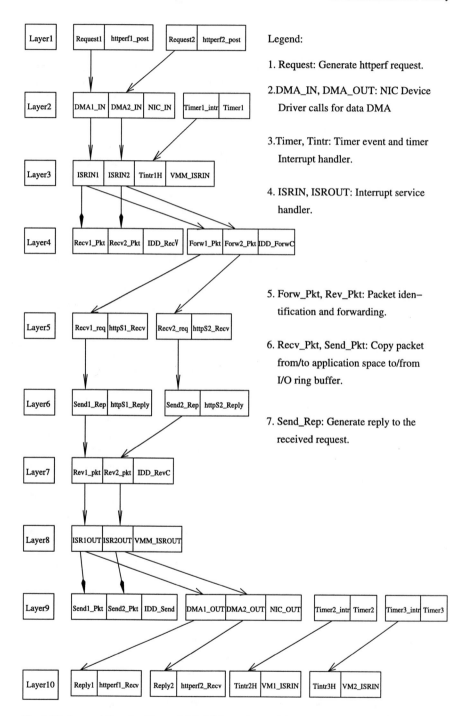

**Fig. 2.13** Layered Queuing Network Model for end-to-end *httperf* benchmark on Xen server

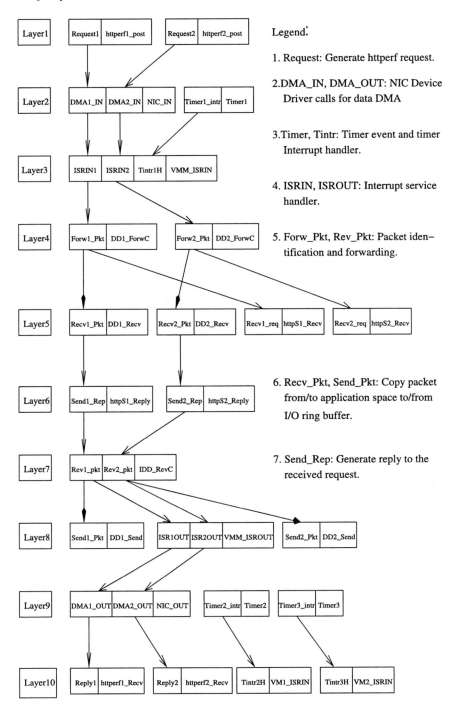

**Fig. 2.14** Layered Queuing Network Model for end-to-end *httperf* benchmark on proposed I/O virtualized server

the achievable throughput when compared to actual implementation. But the deviation is well within 10% of the observed values, as reported in [29, 30]. This makes LQN models very useful in evaluating end-to-end architectures.

3. One element that is incorporated in the LQN model and not shown in the workflow is the system timer interrupt using the server element "Timer." This element is introduced in the LQN to account for the queuing delays accrued, while the OS is handling timer interrupts. For generating the service time of the interrupt handler, a significantly small delay is used. This value is currently set randomly for want of standard tools to profile kernel procedures.

4. All entries in the LQN model that represent hardware functions are set with a significantly small delay as the service time.

Further details on generating of the LQN models and validating the models against experimental data for this benchmark are discussed in [29, 30].

# References

1. Goldberg, R.P.: Survey of virtual machine research. IEEE Comput. **7**(6), 34–45 (1974)
2. Jones, R.A.: Netperf: a network performance benchmark revision 2.0. Technical Report, Information Networks Division, Hewlett-Packard Company (1993). Available online: http://ci.nii.ac.jp/naid/10000088072/en/. Cited 30 April 2010
3. RADS Carleton Univ.: Layered Queueing Network Solver software package (1995). Available online: http://www.sce.carleton.ca/rads/lqns. Cited 30 April 2010
4. Rolia, J.A., Sevcik, K.C.: The method of layers. IEEE Trans. Softw. Eng. **21**(8), 689–700 (1995)
5. Woodside, C.M., Neilson, J.E., Petriu, D.C., Majumdar, S.: The stochastic rendezvous network model for performance of synchronous client-server-like distributed software. IEEE Trans. Comput. **44**(1), 20–34 (1995)
6. Kaashoek, M.F., et al.: Application performance and flexibility on exokernel systems. In: 16th ACM SOSP, pp. 52–65 (1997)
7. Verghese, B., Gupta, A., Rosenblum, M.: Performance isolation: sharing and isolation in shared-memory multiprocessors. ACM SIGPLAN Not. **19**, 181–192 (1998)
8. Mosberger, D., Jin, T.: httperf: a tool for measuring web server performance. In: ACM Workshop on Internet Server Performance, pp. 59–67 (1998)
9. Sugerman, J., Venkatachalam, G., Lim, B.: Virtualizing I/O devices on VMware workstation's hosted virtual machine monitor. In: Proceedings of the USENIX Annual Technical Conference, pp. 1–14 (2001)
10. Welsh, M., Culler, D.: Virtualization considered harmful OS design directions for well-conditioned services. In: Hot Topics in OS 8th Workshop, pp. 139–144 (2001)
11. Barham, P., Dragovic, B., Fraser, K., Hand, S., Harris, T., Ho, A., Neugebauer, R., Pratt, I., Warfield, A.: Xen and the art of virtualization. In: 19th ACM SIGOPS, pp. 164–177 (2003)
12. Fraser, K., Hand, S., Neugebauer, R., Pratt, I., Wareld, A., Williamson, M.: Safe hardware access with the Xen virtual machine monitor. In: 1st Workshop on OASIS (2004)
13. The Globus Resource Specification Language RSL v1.0 (2004). Available online: http://www-fp.globus.org/gram/rsl_spec1.html. Cited 30 April 2010
14. Menon, Santos, J.R., Turner, Y., Janakiraman, G.J., Zwaenepoel, W.: Diagnosing performance overheads in the Xen virtual machine environment. In: Proceedings of the ACM/USENIX Conference on Virtual Execution Environments, pp. 13–23 (2005)
15. Vmware (2005) Vmware ESX Server 2—architecture and performance implications (2005). Available online: http://www.vmware.com/pdf/esx2_performance_implications.pdf. Cited 30 April 2010

16. Raj, H., Schwan, K.: Implementing a scalable selfvirtualizing network interface on a multi-core platform. In: Workshop on the Interaction Between Operating Systems and Computer Architecture (2005)
17. Gupta, D., Cherkasova, L., Gardner, R., Vahdat, A.: Enforcing performance isolation across virtual machines in Xen. Lect. Notes Comput. Sci. **4290**, 342–362 (2006)
18. Intel Virtualization Technology for Directed-I/O (2006). Available online: www.intel.com/technology/itj/2006/v10i3/2-io/7-conclusion.htm. Cited 30 April 2010
19. Liu, J., Huang, W., Abali, B., Panda, D.K.: High performance VMMbypass I/O in virtual machines. In: Proceedings of the USENIX Annual Technical Conference, pp. 3–3 (2006)
20. Menon, Cox, A.L., Zwaenepoel, W.: Optimizing network virtualization in Xen. In: Proceedings of the USENIX Annual Technical Conference, pp. 2–2 (2006)
21. PCI-SIG IOV Specification (2006). Available online: http://www.pcisig.com/specifications/iov. Cited 30 April 2010
22. Santos, J.R., Janakiraman, G., Turner, Y., Pratt, I.: Netchannel 2: optimizing network performance. In: Xen Summit Talk (2007)
23. Willmann, P., Shafer, J., Carr, D., Menon, A., Rixner, S., Cox, A.L., Zwaenepoel, W.: Concurrent direct network access for virtual machine monitors. In: Proceedings of the International Symposium on High-Performance Computer Architecture, pp. 306–317 (2007)
24. Nesbit, K.J., Moreto, M., Cazorla, F.J., Ramirez, A., Valero, M., Smith, J.E.: Multicore resource management. IEEE Micro **28**(3), 6–16 (2008). Special Issue on Interaction of Computer Architecture and Operating System in the Manycore Era
25. Neterion (2008). Available online: http://www.neterion.com/. Cited 30 April 2010
26. Netxen (2008). Available online: http://www.netxen.com/. Cited 30 April 2010
27. Rixner, S.: Breaking the performance barrier: shared I/O in virtualization platforms has come a long way but performance concerns remain. ACM Queue **6**(1), 36 (2008)
28. Sun Microsystems: CrossBow Network Virtualization and Resource Control (2008). Available online: http://www.opensolaris.org/os/community/networking/crossbow_sunlabs_ext.pdf. Cited 30 April 2010
29. Lakshmi, J., Nandy, S.K.: Modeling Architecture-OS interactions using layered queuing network models. In: International Conference Proceedings of HPC Asia, pp. 382–389 (2009)
30. Lakshmi, J., Nandy, S.K.: I/O device virtualization in multi-core era, a QoS perspective. In: Workshop on Grids, Clouds and Virtualization, Conference on Grids and Pervasive Computing, pp. 128–135 (2009)
31. Kim, H., Lim, H., Jeong, J., Jo, H., Lee, J.: Task-aware virtual machine scheduling for I/O performance. In: Proceedings of ACM SIGPLAN/SIGOPS International Conference on Virtual Execution Environments, pp. 101–110 (2009)
32. Weng, C., Wang, Z., Li, M., Lu, X.: The hybrid scheduling framework for virtual machine systems. In: Proceedings of ACM SIGPLAN/SIGOPS International Conference on Virtual Execution Environments, pp. 111–120 (2009)
33. Solarflare Communications (2009). Available online: http://www.solarflare.com/. Cited 30 April 2010
34. Linux Advanced routing and Traffic control HowTo. Available online: http://lartc.org/howto/index.html. Cited 30 April 2010
35. Lakshmi, J., Nandy, S.K.: I/O virtualization architecture for security. In: IEEE Proceedings of International Workshop on Virtualization Technology (2010)

# Chapter 3
# Architectures for Enhancing Grid Infrastructures with Cloud Computing

**Eduardo Huedo, Rafael Moreno-Vozmediano, Rubén S. Montero, and Ignacio M. Llorente**

**Abstract**  Grid and Cloud Computing models pursue the same objective of constructing large-scale distributed infrastructures, although focusing on complementary aspects. While grid focuses on federating resources and fostering collaboration, cloud focuses on flexibility and on-demand provisioning of virtualized resources. Due to their complementarity, it is clear that both models, or at least some of their concepts and techniques, will coexist and cooperate in existing and future e-infrastructures. This chapter shows how Cloud Computing will help both to overcome many of the barriers to grid adoption and to enhance the management, functionality, suitability, energy efficiency, and utilization of production grid infrastructures.

## 3.1 Introduction

Grid infrastructures offer common APIs and service interfaces that make it possible to take advantage of distributed resources without having to modify applications for each site. However, this uniformity unfortunately does not extend to the underlying computing resources, where users are exposed to significant heterogeneities in the computing environment, complicating applications and increasing failure rates.

On the other hand, virtualization technologies have matured rapidly over the last few years, providing a mechanism for offering customized, uniform environments

E. Huedo (✉) · R. Moreno-Vozmediano · R.S. Montero · I.M. Llorente
Universidad Complutense de Madrid, 28040 Madrid, Spain
e-mail: ehuedo@fdi.ucm.es

R. Moreno-Vozmediano
e-mail: rmoreno@dacya.ucm.es

R.S. Montero
e-mail: rubensm@dacya.ucm.es

I.M. Llorente
e-mail: llorente@dacya.ucm.es

M. Cafaro, G. Aloisio (eds.), *Grids, Clouds and Virtualization,*
Computer Communications and Networks,
DOI 10.1007/978-0-85729-049-6_3, © Springer-Verlag London Limited 2011

for users. This additional flexibility comes with negligible costs in terms of processing power, network bandwidth, and disk I/O in modern systems. Using grid technologies combined with virtualization will allow the grid to provide users with a homogeneous computing environment, simplifying applications and reducing failures.

In parallel with the increasing maturity of virtualization, Cloud Computing technologies have emerged. These technologies allow users to dynamically allocate computing resources and to specify the characteristics for the allocated resources. The fusion of cloud and grid technologies would provide a more dynamic and flexible computing environment for grid application developers.

In computing, a "cloud" usually refers to an "Infrastructure-as-a-Service" (IaaS) cloud, such as Amazon EC2, where IT infrastructure is deployed in the provider's datacenter in the form of Virtual Machines (VM). Cloud computing enables the deployment of an entire IT infrastructure without the associated capital costs, paying only for the used capacity. This new resource provisioning paradigm has been introduced to better respond to changing computing demands, allowing the increase or decrease of capacity in order to meet peak or fluctuating service demands.

With the growing popularity of IaaS clouds, an ecosystem of tools and technologies is emerging that can be used to transform an organization's existing infrastructure into a *private* cloud, so providing a dynamic and flexible private infrastructure to run virtualized service workloads. Private clouds can also support a *hybrid* cloud model by supplementing local infrastructure with computing capacity from an external *public* cloud. A private/hybrid cloud can allow remote access to its resources over the Internet using remote interfaces, such as the web service interfaces used in Amazon EC2, thus making it also a public cloud.

Cloud and virtualization technologies also offer other benefits to administrators of resource centers, such as the migration of live services for load balancing or the deployment of redundant servers. Reduced costs for managing resources immediately benefits users in freeing money for additional computing resources or in having better user support from administrators.

This chapter shows how cloud technology could enhance existing and future grid infrastructures. The structure of this chapter is as follows. Section 3.2 introduces techniques to enhance grid infrastructures with Cloud Computing. In particular, we envisage the virtualization of grid sites, the delivery of IaaS in grid sites, the cloud scale-out of grid sites, and the federation of grids and clouds. These approaches will be explained in Sects. 3.3 to 3.6. Finally, Sect. 3.7 provides some conclusions.

## 3.2 Grid Infrastructure Enhancement with Cloud Computing

In the last decade we have witnessed the consolidation of several transcontinental grid infrastructures that have achieved unseen levels of resource sharing. In spite of this success, current grids suffer from several obstacles that limit their efficiency, namely:

- An increase in the cost and length of the application development and porting cycle. New applications have to be tested in a great variety of environments where the developers have limited configuration capabilities.
- A limitation on the effective number of resources available to each application. Usually, a Virtual Organization (VO) requires a specific software configuration, so an application can be only executed on those sites that support the associated VO. Moreover, the resources devoted to each VO within a site are usually static and cannot be adapted to the VO's workload.
- An increase in the operational cost of the infrastructure. The deployment, maintenance, and distribution of different configurations requires specialized, time consuming, and error prone procedures. Even worse, new organizations joining a grid infrastructure need to install and configure an ever-growing middleware stack.

Grid infrastructures can overcome these limitations and can be enhanced in several ways through the use of Cloud Computing concepts. However, we must keep in mind that grid technologies, policies, and procedures are the result of many years of research, development, and operation. Therefore, we should propose evolutionary, and not revolutionary, steps in the development of better research infrastructures. The approaches we envision are the following:

- Virtualization of grid sites. The integration of private cloud technologies and services, especially virtualization, into existing grid infrastructures would enhance failover and redundancy solutions, and permit machine migration for flexible load balancing and energy efficiency. Virtualization of a grid site would also allow the dynamic provisioning of worker nodes to address the demands of different user communities. This approach will address several needs from resource providers in existing grid infrastructures, being fully transparent to grid application communities, while users would benefit indirectly via the improved stability, reliability, and robustness of the infrastructure.
- IaaS delivery in grid sites. The provision of infrastructure using cloud-like delivery paradigms in addition to existing grid services will address the emerging IaaS cloud-like usage patterns from several user communities. Public cloud interfaces would provide an alternative access to grid site resources to support the execution of any application encapsulated in a VM image. The new interfaces will complement existing grid services, providing a new way to access to the same underlying grid site infrastructure without replacing the grid functionality. In this case, new grid user communities and industrial users would benefit from this innovation in the resource provisioning model of grid sites.
- Cloud scale-out of grid sites. Using hybrid cloud technologies would additionally support "elastic" grid sites able to expand available computing resources in the local cloud to meet peak demands using remote cloud providers. Again, this approach will ease capacity planning for resource providers in existing grid infrastructures, allowing them a quick reaction to peak loads, and still being fully transparent to grid application communities. Users would benefit indirectly via the on-demand increasing capacity of the infrastructure.

- Federation of grids and clouds. Virtual clusters or grids, as well as individual nodes, can be deployed in public clouds to be accessed from current grid infrastructures using a metascheduler or broker. This way, we would have a federated infrastructure with physical resources statically provisioned (with shared access) from grids complemented with virtual resources dynamically provisioned (with exclusive access) from clouds when needed to meet a given SLA (Service Level Agreement). This technique allows the provision of cloud resources from current grid infrastructures without any change.

The above approaches, based on cloud and virtualization techniques, would provide flexibility, energy efficiency, and elasticity to grid sites, and would maximize the utility of grid resources for existing user communities. The first three approaches can be seen as a natural evolution of a grid site, first to become a private cloud, then a public cloud, and finally an hybrid cloud. The fourth approach aims to federate conventional grid sites and cloud resources. These four approaches will be elaborated more in the next sections.

The RESERVOIR[1] and EGEE[2] projects are working together to explore how the institutes providing computing resources to EGEE could benefit from adopting private and hybrid cloud models to provide resources [17]. In particular, the use of a cloud-like provisioning model will allow one to easily meet the changing needs of the grid users, from scaling up services to meeting peak loads and improving redundancy or to changing the resources provided to run particular applications. In the context of this collaboration, the StratusLab initiative[3] was created, as an informal collaboration framework, to evaluate the maturity of existing cloud and virtualization technologies and services to enhance production grid infrastructures, and to promote the benefits of virtualization and cloud for the grid community [18].

## 3.3 Virtualization of Grid Sites

The pattern of resource demand of the computing community is strongly variable, so making quite difficult to estimate the resource demands. Resource providers need infrastructure solutions, controlled by site administrators, to meet peak demands in their clusters. The usual grid answer is to share between disciplines to smooth out the peaks.

Moreover, in order to be responsive to user requests, the grid must be able to more easily allocate and reallocate its resources. This is currently a problem with the current grid implementation, constrained by its technological choice. Most grid users require a carefully setup environment for their applications to run—e.g., operating system, libraries, applications, or shared file system—which forces the system

---

[1] www.reservoir-fp7.eu

[2] www.eu-egee.org

[3] www.stratuslab.org

administrator to comply with these requirements in order, for a given VO, to successfully run on their resources. In the past, this has also constrained the users to be conservative in their choices of runtime environment, in order to avoid difficult negotiations with the resource owners as time goes on.

In any case, due to the heterogeneity in resource configurations, resources are only useful for a subset of the full user community. Therefore, heterogeneity reduces the opportunities for sharing because underused resources from one community cannot be used to meet peak resource demands from another. Moreover, existing grids suffer from the lack of ability to adapt to the exact requirements of the end-users. Learning from cloud technologies and virtualization, we can now consider a new mode of operation, which completely removes the point of friction between users, looking for a fully customized environment, and administrator, looking for a fully homogeneous one.

Several alternatives have been explored in the past to solve this. For example, the SoftEnv project is a software environment configuration system that allows the users to define the applications and libraries they need [29]. Another common solution is the use of a custom software stack on top of the existing middleware layer, usually referred as pilot-jobs. For example, MyCluster creates a Condor or Sun Grid Engine (SGE) cluster on top of TeraGrid services [31]; and similarly over other middleware we may cite DIRAC [30], glideinWMS [28], or PanDa [22]. Additionally, several projects have investigated the partitioning of a distributed infrastructure to dynamically provide customized independent clusters. For example, COD (Cluster On Demand) is a cluster management software which dynamically allocates servers from a common pool to multiple virtual clusters [6].

However, the most promising technology to provide VOs with custom execution environments is virtualization. The dramatic performance improvements in hypervisor technologies made possible to experiment with VMs as basic building blocks for computational platforms. In fact, several studies reveal that the virtualization layer has no significant impact on the performance of memory and CPU-intensive applications for HPC clusters [32, 33] or grids[12].

Virtualization offers an attractive solution since it completely separates the host machine (under the control of the system administrators of the resource provider) and the VM running the user operating system. Translated to grid sites, it means that the system administrators remain in full control of their infrastructure, following their own update and upgrade schedule, and remaining free to setup their resource environment as they see fit, as long as they of course comply with the minimum requirements for running VMs. Meanwhile, grid users are now free to compose their VMs exactly as they want it. While there are a few constraints on the VMs depending on the virtualization technology supported, these remain minor compared to the significant benefit that virtualization offers.

The first works in this area integrated resource management systems with VM to provide custom execution environments on a per-job basis. For example, Dynamic Virtual Clustering and XGE for MOAB and SGE job managers, respectively [8, 10]. These approaches only overcome the configuration limitation of physical resources because VMs are bounded to a given resource and only exist during job execution. A similar approach has been implemented at grid level using the Grid-Way Metascheduler [26]. GridWay[4] allows the definition of an optional phase before the actual execution phase to perform advanced job configuration routines. In this phase, a user-defined program (pre-wrapper), executed on the cluster front-end, checks the availability of the requested VM image in the cluster node, transferring it from a GridFTP repository if needed. Then, in the execution phase, another program (wrapper) is executed on a worker node of the cluster. This program starts or restores the VM and waits for its activation by periodically probing its services. When the VM is ready, the program copies all the input files needed to the VM and executes the user program. When this program exits, output files are copied to the client host, and the VM is shut down (or suspended to disk to be recovered later). This strategy does not require additional middleware to be deployed and is not tied to a given virtualization technology. However, since the underlying local resource management system is not aware of the nature of the job itself, some of the potential benefits offered by the virtualization technology (e.g., server consolidation) are not fully exploited.

More general approaches involve the use of VMs as workload units, which implies the change in paradigm from building grids out of physical resources to virtualized ones. For example, the VIOLIN project proposes a novel alternative to application-level overlays based on virtual and isolated networks created on top of an overlay infrastructure. Also, the VMPlant service provides the automated configuration and creation of VMs that can be subsequently be cloned and instantiated to provide homogeneous execution environments across distributed grid resources [15]. On the other hand, the In-VIGO project adds some virtualization layers to the classical grid model, to enable the creation of dynamic pools of virtual resources for application-specific grid computing [1]. Finally, several studies have explored the use of VMs to provide custom (VO-specific) cluster environments for grid computing. In this case, the clusters are usually completely build up of virtualized resources, as in the Globus Nimbus project [11] or the Virtual Organization Clusters (VOC) [21].

The virtualization of a grid site using private cloud technologies, as Fig. 3.1 shows, would enable it to meet the changing needs of the users, by adapting and customizing the infrastructure to offer the services required by different application communities. This would allow the grid infrastructure to maximize the utility of the resources, better supporting sharing of resources between communities. For this to work, the right VM has to be instantiated to run a given grid job. Since the cloud API already supports the ability to dynamically instantiate a number of VMs, from an existing store of virtual images, it is possible to manage user-defined virtual images, once the grid API allows such information to be transmitted from the user to the resource manager.

Previous works also highlight that the use of virtualization in grid environments can greatly improve the efficiency, flexibility, and sustainability of current productions grids. For example, private cloud technologies would also help sites to dynam-

---

[4]www.gridway.org

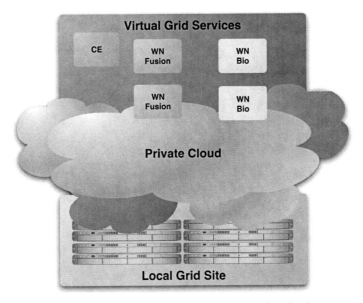

**Fig. 3.1** Virtualization of a grid site to address quick deployment of applications

ically consolidate grid services on a lower number of physical resources, reducing the number of active physical systems and thus the administrative effort, power, and cooling required. It is worth noting that energy efficiency is a critical issue for research infrastructures today [3].

Summing up, grids can take advantage of virtualization, not only by extending the classical benefits of VMs for constructing cluster, e.g., consolidation or rapid provisioning of resources [9, 20, 23], but also by obtaining grid-specific benefits, e.g., support to multiple VOs, isolation of workloads, and the encapsulation of services [2].

## 3.4 IaaS Delivery in Grid Sites

Many resource providers are interested in using their physical infrastructure to perform other tasks apart from grid service execution, e.g., development of new codes, teaching, creating an internal computing cluster, etc. Moreover, they would like to decide the fraction of resources available via the grid. Resource providers need administrative solutions to easily partition and isolate different clustered services running on the same infrastructure. In fact some resource providers are deploying private cloud facilities in order to support various activities, the grid infrastructure being one of them. The virtualization of the grid site using private cloud technologies will allow organizations to execute multiple virtualized clustered services on the same physical cluster, dynamically allocating different capacity to the services. Because the same physical infrastructure could be shared by different services, pri-

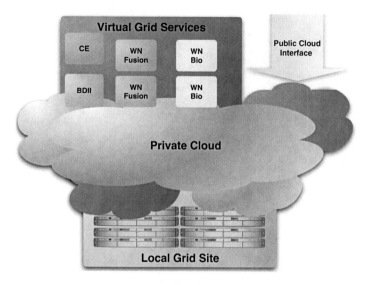

**Fig. 3.2** Offering infrastructure as a service

vate cloud computing will increase the number of resources provided by existing grid sites and reduce administration effort.

Once a grid site is virtualized, as explained in the previous section, the provision of infrastructure cloud interfaces, as shown on Fig. 3.2, would provide an alternate, complementary access to grid site resources and would support the execution of any application encapsulated in a VM image. Using virtualization would increase the number of users by making the hardware useful to a wider range of applications. Moreover, this new functionality would reduce the required manpower in application porting and would attract the science user communities and industrial users that have embraced the cloud computing provisioning model.

This will allow grid sites to turn their site into a public cloud. This way, current and future sites will be equipped with a much more powerful and flexible mean of giving access to their resources to a wider range of users, while not compromising important aspects in data centre managements, such as security, traceability, and auditing.

It is important that this cloud API is introduced in a harmonious way with respect to current grid APIs. It is also important that current users of the grid find a sensible migration path to this new way of accessing resources. At the same time, it is equally important that new users find this cloud API to grid resources as convenient and straightforward as possible. In other words, this is an opportunity to streamline the usage of the grid.

Having an infrastructure that combines both technologies allows it to serve the maximum number of users, including traditional grid users with computational resources to federate, as we will show in Sect. 3.6, and potential new communities that have financial resources to pay for resource utilization, but no resources of their own. This flexibility allows users to structure their applications in a way that is most efficient for them without having to deploy separate resources for each type of infrastructure. It would also allow them to take advantage of commercial providers.

The OpenNebula[5] virtual infrastructure manager provides cloud interfaces like EC2 Query and the OCCI (Open Cloud Computing Interface) standard. Therefore, grid sites virtualized using this technology as virtual infrastructure manager are ready to offer IaaS with very little effort.

## 3.5   Cloud Scale-Out of Grid Sites

Hybrid cloud technologies provide an additional method to deal with peak loads, enhancing the system administrator's control over the system. Hybrid cloud computing is also the bridge between existing grid infrastructures and new emerging commercial and science infrastructures based on the cloud model.

Few studies have explored this hybrid cloud model. For example, the VioCluster project enables to dynamically adjust the capacity of a computing cluster by sharing resources between peer domains [27]. Also, as we will show later, the Open-Nebula allows the creation of clusters combining physical, virtualized, and cloud resources [16]. Finally, important work has already taken place by the open StratusLab collaboration to evaluate the feasibility of running an entire grid site in the Amazon cloud [19], dispensing with any local infrastructure.

A grid site can support a VO beyond its physical resource capabilities by scaling-out to an external cloud provider, such as commercial cloud providers, as it is shown on Fig. 3.3. While, in the case of a commercial cloud provider, the site would have to pay for the resources scaled-out, this means that unprecedented flexibility is provided to grid site administrator. However, control must be put in place to avoid abuse.

This possibility could have a significant impact in reducing infrastructure expenditure since resource owners will no longer have to size their infrastructure for the worst-case scenario (peak demand) but for average expected demand. This is possible since the peak demand would simply be scaled-out to third party cloud provider. Over time, as grid sites could also offer a cloud API, as we explained in the previous section, they in return might become resource providers allowing other sites to utilize their excess capacity, if any at a give time. The idea here is that, instead of interchanging jobs, grid sites would interchange resources, encapsulated in VMs. Therefore, this model could in time increase the fluidity of resource allocation and improve resource utilization on a global scale.

Figure 3.4 shows a possible implementation of a hybrid cloud using OpenNebula. OpenNebula provides a uniform management layer regardless of the underlying virtualization technology. In this way, OpenNebula can be easily integrated with cloud services by using a specific Amazon EC2 plug-in. The EC2 plug-in then converts the general requests made by OpenNebula core, such as deploy or shutdown, using the EC2 API.

---

[5] www.opennebula.org

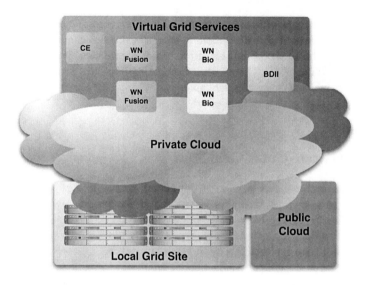

**Fig. 3.3** Combining a grid site with cloud resources to meet a peak demand

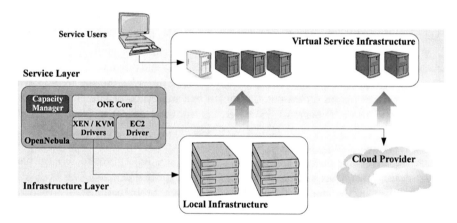

**Fig. 3.4** Hybrid cloud with OpenNebula

## 3.6 Federation of Grids and Clouds

In the recent years there have been a lot of efforts aimed at providing interoperation between grid middlewares to allow the federation of grid infrastructures [24]. This grid interoperation and federation techniques can be also extended to cloud infrastructures.

A set of nodes, possibly grouped in a virtual cluster or a grid site (as explained in the previous section), can be deployed in public clouds to be accessed from current grid infrastructures using a metascheduler or broker, i.e., using Grid Computing concepts, creating a federated infrastructure with physical resources from grids

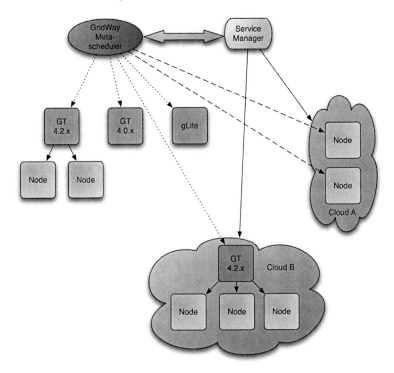

**Fig. 3.5** Federating grids and clouds

complemented with dynamically provisioned virtual resources from clouds when needed, for example, to face peak demands or to meet a given SLA (Service Level Agreement). Grid resources would be shared, while cloud resources would be exclusive, at a given price. This technique allows the provision of cloud resources and their use from current grid infrastructures without any change.

Some architectures have been proposed for this. For example, the InterGrid system uses VMs as building blocks to construct execution environments that span multiple computing sites [7]. Such environments can be created by deploying VMs on different types of resources, like local data centers, grid infrastructures or cloud providers. InterGrid uses OpenNebula as a component for deploying VMs on a local infrastructure. Also, Fig. 3.5 sketches an architecture of a grid infrastructure that can be flexibly built, incorporating new resources temporarily in an automatic fashion to satisfy heavy demands [4]. Moreover, if there is one specific service which is suffering from the peak demand, the system can decide to increase the number of nodes prepared to satisfy such a service. As can be seen, this approach is similar to the cloud scale-out approach presented in the previous section, but now the federation is based on Grid Computing concepts, instead of hybrid Cloud Computing. In addition, we think that this architecture is more natural when using commercial public clouds, since the end user, and not the site administrator, is responsible for the allocation of virtual resources (and will receive the invoice).

One of the building blocks of the architecture presented in Fig. 3.5 is again the GridWay Metascheduler. The flexible architecture of this metascheduler allows the use of adapters (called Middleware Access Drivers, or MADs, in GridWay's terminology), that enable access to different production grid infrastructures [14]. Moreover, it also provides SSH adapters, so access to single nodes can be achieved with decreased overhead, avoiding the need to have installed and configured in the nodes any grid software as, for instance, the Globus Toolkit. Also, the GridWay metascheduler features mechanisms to dynamically discover new resources, and it is able to detect and recover from any of the grid elements failure, outage, or saturation conditions [13]. Moreover, the metascheduler can handle differences in latencies and performance of resources.

The principal component of the proposed architecture for dynamic provisioning is the Service Manager. This component is used to monitor the GridWay Metascheduler, and when the load of the system excesses a threshold, detected using heuristics, it is responsible to grow the available grid infrastructure using specific adapters to access different cloud providers. Of course, this component is also responsible to shrink the infrastructure when the load decreases. Clearly, resource provision heuristics should take into account both QoS (Quality of Service) and budget constraints [5] of the end user.

Grid infrastructure growth can be accomplished in two ways. The first one is by requesting a number of single hosts. This corresponds to the use of cloud A in Fig. 3.5. Therefore, this mode adds one single computing resource to the grid infrastructure that will be accessible through SSH to perform job execution, as GridWay already has such SSH drivers. In this way, machines from cloud providers can be used out-of-the-box, with little to nonconfiguration needed, since basically SSH access is the only requirement.

Another possibility is to deploy a fully virtualized cluster, with a front-end controlling a number of slave nodes. This front-end can then be enrolled to the existing grid infrastructure, adding its capacity [25]. This corresponds to the use of cloud B in the figure. In this second model, negotiation with the cloud provider will grant access to a virtual cluster, accessible through Globus GRAM and controlled by a local resource manager like, for example, PBS or SGE. This cluster will then be added to the federated grid infrastructure the same way as any other physical sites. Future work is planned to enrich the flexibility of the grid infrastructure by removing the GRAM layer, enabling GridWay to access the cluster by talking directly to the local resource manager, using the DRMAA (Distributed Resource Management Application API) standard.

## 3.7 Conclusions

Cloud Computing has emerged as a very promising paradigm to simplify and improve the management of current IT infrastructures of any kind. Clouds, in their IaaS form, have opened up avenues in this area to ease the maintenance, operation, and use of grid sites, and to explore new resource sharing models that could simplify in some cases the porting and development of grid applications. The first works about the joint use of clouds and grids are exploring two main approaches, namely:

- The use of virtualization and cloud techniques as an effective way to provide grid users with custom execution environments. So the same grid site can easily support VOs with different (or even conflicting) configurations. Moreover, grid sites would benefit from improved flexibility, reliability, and efficiency.
- The access to grid resources in a cloud-way. So, the users will access "raw" computing capacity bypassing the classical grid middleware stack. This approach is also being considered as a natural way to attract business users to our current e-infrastructures.

The potential benefits that cloud and virtualization technologies can bring to current e-infrastructures require of a common framework that bridge grid and cloud computing models. Various solutions have been proposed to take advantage of these new technologies in a grid environment, from its direct application to encapsulate the execution of each job to the advance provisioning of virtual clusters.

Grid and cloud technologies address fundamentally different aspects of distributed computing. Grid technology focuses on federation of resources, uniform APIs, common authorization mechanisms, and sharing of resources while cloud computing focuses on dynamic, easy access to resources. Grid site management can be enormously simplified with cloud technologies. But, similarly, cloud resources can be improved by using the common grid authorization mechanisms and movement of files (like VM images) between resources. Because of these complementarities, these technologies will coexist for the foreseeable future, and platforms combining them will offer their users a better service.

**Acknowledgements** This research was supported by Consejería de Educación de la Comunidad de Madrid, Fondo Europeo de Desarrollo Regional (FEDER) and Fondo Social Europeo (FSE), through MEDIANET Research Program S2009/TIC-1468, by Ministerio de Ciencia e Innovación, through the research grant TIN2009-07146, and by the European Union through the StratusLab contract number RI-261552.

# References

1. Adabala, S., Chadha, V., Chawla, P., et al.: From virtualized resources to virtual computing grids: the In-VIGO system. Future Gener. Comput. Syst. **21**(6), 896–909 (2005)
2. Begin, M.: An EGEE comparative study: grids and clouds—evolution or revolution. Tech. rep., EGEE-III NA1 (2008). Available at http://edms.cern.ch/file/925013
3. Berl, A., Gelenbe, E., Di Girolamo, M., Giuliani, G., De Meer, H., Dang, M.Q., Pentikousis, K.: Energy-efficient cloud computing. Comput. J. **53**(7), 1045–1051 (2010)
4. Blanco, C.V., Huedo, E., Montero, R.S., Llorente, I.M.: Dynamic provision of computing resources from grid infrastructures and cloud providers. In: Proceedings of the Workshop on Grids, Clouds and Virtualization, in Conjunction with Grid and Pervasive Computing Conference (GPC 2009), pp. 113–120. IEEE Computer Society, Los Alamitos (2009)
5. Buyya, R., Murshed, M.M., Abramson, D., Venugopal, S.: Scheduling parameter sweep applications on global Grids: a deadline and budget constrained cost-time optimization algorithm. Softw. Pract. Exp. **35**(5), 491–512 (2005)

6. Chase, J.S., Irwin, D.E., Grit, L.E., Moore, J.D., Sprenkle, S.E.: Dynamic virtual clusters in a grid site manager. In: Proceedings of the 12th International Symposium on High Performance Distributed Computing (HPDC 2003) (2003)
7. di Costanzo, A., de Assuncao, M., Buyya, R.: Harnessing cloud technologies for a virtualized distributed computing infrastructure. IEEE Internet Comput. **13**(5), 24–33 (2009)
8. Emeneker, W., Jackson, D., Butikofer, J., Stanzione, D.: Dynamic virtual clustering with Xen and Moab. In: Proceedings of the Frontiers of High Performance Computing and Networking, ISPA 2006 Workshops. Lecture Notes in Computer Science, vol. 4331, pp. 440–451. Springer, Berlin (2006)
9. Emeneker, W., Stanzione, D.: Dynamic virtual clustering. IEEE Cluster (2007)
10. Fallenbeck, N., Picht, H., Smith, M., Freisleben, B.: Xen and the art of cluster scheduling. In: Proceedings of the 1st International Workshop on Virtualization Technology in Distributed Computing (VTDC 2006) (2006)
11. Foster, I., Freeman, T., Keahey, K., Scheftner, D., Sotomayor, B., Zhang, X.: Virtual clusters for grid communities. In: Proceedings of the 6th IEEE International Symposium on Cluster Computing and the Grid (CCGrid 2006) (2006)
12. Gilbert, L., Tseng, J., Newman, R., Iqbal, S., Pepper, R., Celebioglu, O., Hsieh, J., Mashayekhi, V., Cobban, M.: Implications of virtualization on grids for high energy physics applications. J. Parallel Distrib. Comput. **66**(7), 922–930 (2006)
13. Huedo, E., Montero, R.S., Llorente, I.M.: The GridWay framework for adaptive scheduling and execution on grids. Scalable Comput. Pract. Exp. **6**, 1–8 (2005)
14. Huedo, E., Montero, R.S., Llorente, I.M.: A modular meta-scheduling architecture for interfacing with pre-WS and WS grid resource management services. Future Gener. Comput. Syst. **23**(2), 252–261 (2007)
15. Krsul, I., Ganguly, A., Zhang, J., Fortes, J.A.B., Figueiredo, R.J.: VM-Plants: Providing and managing virtual machine execution environments for grid computing. In: Proceedings of the 2004 ACM/IEEE Conference on Supercomputing (2004)
16. Llorente, I.M., Moreno-Vozmediano, R., Montero, R.S.: Cloud computing for on-demand grid resource provisioning. In: Proceedings of the High Performance Computing Workshop 2008, High Speed and Large Scale Scientific Computing. Advances in Parallel Computing, vol. 18, pp. 177–191. IOS, Amsterdam (2009)
17. Llorente, I.M., Newhouse, S.: Collaboration between the EGEE and RESERVOIR projects. In: EGEE 2009 Conference (2009)
18. Loomis, C.: StratusLab—enhancing grid infrastructures with cloud computing. In: EGEE 2009 Conference (2009)
19. Loomis, C., Begin, M., Floros, V., Llorente, I.M., Montero, R.S.: Operating a grid site in the cloud. In: 4th EGEE User Forum/OGF 25 and OGF Europe's 2nd International Event (2009)
20. Moreno-Vozmediano, R., Montero, R.S., Llorente, I.M.: Elastic management of cluster-based services in the cloud. In: Proceedings of the 1st Workshop on Automated Control for Datacenters and Clouds (ACDC 2009), in Conjunction with 6th International Conference on Autonomic Computing and Communications (ICAC 2009), pp. 19–24. ACM, New York (2009)
21. Murphy, M., Kagey, B., Fenn, M., Goasguen, S.: Dynamic provisioning of virtual organization clusters. In: Proceedings of the 9th IEEE International Symposium on Cluster Computing and the Grid (CCGrid 2009) (2009)
22. Nilsson, P.: Experience from a pilot based system for ATLAS. J. Phys. Conf. Ser. **119**(6), 062038 (2008)
23. Nishimura, H., Maruyama, N., Matsuoka, S.: Virtual clusters on the fly—fast, scalable, and flexible installation. In: Proceedings of the 7th IEEE International Symposium on Cluster Computing and the Grid (CCGRID 2007) (2007)
24. Riedel, M., Laure, E., et al.: Interoperation of world-wide production e-Science infrastructures. Concurr. Comput. Pract. Exp. **21**(8), 961–990 (2009)
25. Rodriguez, M., Tapiador, D., Fontan, J., Huedo, E., Montero, R.S., Llorente, I.M.: Dynamic provisioning of virtual clusters for grid computing. In: Proceedings of the 3rd Workshop on

Virtualization in High-Performance Cluster and Grid Computing (VHPC 2008), in Conjunction with Euro-Par 2008. Lecture Notes in Computer Science, vol. 5415, pp. 23–32. Springer, Berlin (2009)

26. Rubio-Montero, A., Huedo, E., Montero, R., Llorente, I.: Management of virtual machines on Globus Grids using GridWay. In: High Performance Grid Computing Workshop (HPGC 2007), in Conjunction with 21th International Parallel and Distributed Processing Symposium (IPDPS 2007), pp. 1–7 (2007)

27. Ruth, P., McGachey, P., Xu, D.: Viocluster: virtualization for dynamic computational domains. In: 2005 IEEE International Conference on Cluster Computing (2005)

28. Sfiligoi, I.: glideinWMS—a generic pilot-based workload management system. J. Phys. Conf. Ser. **119**(6), 062,044 (2008)

29. Teragrid User Support: Managing Your Software Environment. Available at https://www. teragrid.org/web/user-support/environment. Accessed April 2010

30. Tsaregorodtsev, A., Garonne, V., Stokes-Rees, I.: DIRAC: a scalable lightweight architecture for high throughput computing. In: Proceedings of the 5th IEEE/ACM International Workshop on Grid Computing (GRID'04), pp. 19–25 (2004)

31. Walker, E., Gardner, J.P., Litvin, V., Turner, E.: Creating personal adaptive clusters for managing scientific jobs in a distributed computing environment. In: Proceedings of the IEEE Workshop on Challenges of Large Applications in Distributed Environments (CLADE 2006) (2006)

32. Youseff, L., Seymour, K., You, H., Dongarra, J., Wolski, R.: The impact of paravirtualized memory hierarchy on linear algebra computational kernels and software. In: Proceedings of the High Performance Distributed Computing (HPDC) (2008)

33. Youseff, L., Wolski, R., Gorda, B., Krintz, C.: Paravirtualization for HPC systems. In: Proceedings of the Workshop on XEN in HPC Cluster and Grid Computing Environments (XHPC), in Conjunction with International Symposium on Parallel and Distributed Processing and Application (ISPA 2006) (2006)

# Chapter 4
# Scientific Workflows in the Cloud

Gideon Juve and Ewa Deelman

**Abstract** The development of cloud computing has generated significant interest in the scientific computing community. In this chapter we consider the impact of cloud computing on scientific workflow applications. We examine the benefits and drawbacks of cloud computing for workflows, and argue that the primary benefit of cloud computing is not the economic model it promotes, but rather the technologies it employs and how they enable new features for workflow applications. We describe how clouds can be configured to execute workflow tasks and present a case study that examines the performance and cost of three typical workflow applications on Amazon EC2. Finally, we identify several areas in which existing clouds can be improved and discuss the future of workflows in the cloud.

## 4.1 Introduction

In this chapter we consider the use of cloud computing for scientific workflow applications. Workflows are coarse-grained parallel applications that consist of a series of computational tasks logically connected by data- and control-flow dependencies. They are used to combine several different computational processes into a single coherent whole. Many different types of scientific analysis can be easily expressed as workflows, and, as a result, they are commonly used to model computations in many science disciplines [13]. Using workflow technologies, components developed by different scientists, at different times, for different domains can be used together. Scientific workflows are used for simulation, data analysis, image processing, and many other functions.

G. Juve (✉) · E. Deelman
University of Southern California, Marina del Rey, CA, USA
e-mail: juve@usc.edu

E. Deelman
e-mail: deelman@isi.edu

M. Cafaro, G. Aloisio (eds.), *Grids, Clouds and Virtualization,*
Computer Communications and Networks,
DOI 10.1007/978-0-85729-049-6_4, © Springer-Verlag London Limited 2011

Scientific workflows can range in size from just a few tasks to millions of tasks. For large workflows, it is often desirable to distribute the tasks across many computers in order to complete the work in a reasonable time. As such, workflows often involve distributed computing on clusters, grids [18], and other computational infrastructures. Recently cloud infrastructures are also being evaluated as an execution platform for workflows [24, 28].

Cloud computing represents a new way of thinking about how to deploy and execute scientific workflows. On the one hand, clouds can be thought of as just another platform for executing workflow applications. They support all of the same techniques for workflow management and execution that have been developed for clusters and grids. With very little effort a scientist can deploy a workflow execution environment that mimics the environment they would use on a local cluster or national grid. On the other hand, clouds also provide several features, such as virtualization, that offer new opportunities for making workflow applications easier to deploy, manage, and execute. In this chapter we examine those opportunities and describe how workflows can be deployed in the cloud today.

Many different types of clouds are being developed, both as commercial ventures and in the academic community. For the purposes of this chapter, we are primarily interested in Infrastructure as a Service (IaaS) clouds [3] as these are more immediately useable by workflow applications. Other clouds, such as Platform as a Service (PaaS) and Software as a Service (SaaS) clouds, may provide additional benefits for creating, managing, and executing workflow-based computations, but currently there is a lack of systems developed in this area, and additional research is needed to determine how such systems can be fruitfully combined with workflow technologies.

## 4.2 Workflows in the Cloud

There is some disagreement about what is the killer feature of cloud computing. For many, especially those in the business community, the attractiveness of cloud is due to its utility-based computing model—the idea that someone else manages a set of computational resources and users simply pay to access them. The academic community, however, has had utility computing for quite a long time in the form of campus clusters, high-performance computing centers such as the NCSA [37] and the SDSC [44], and national cyberinfrastructure such as the TeraGrid [48] and the Open Science Grid [38]. Although the availability of commercial clouds may have some impact on the economics of large-scale scientific computing, we do not view economics as the fundamental benefit of cloud computing for science. Instead, we think that clouds provide a multiplicity of benefits that are more technological in nature and that these benefits stem, primarily, from the extensive use of service-oriented architectures and virtualization in clouds. In the following sections we discuss several aspects of cloud computing that are of particular benefit to workflow applications.

## 4.2.1 Provisioning

In grids and clusters scheduling is based on a best-effort model in which a user specifies their computation (the number of resources and amount of time required) and delegates responsibility for allocating resources and scheduling the computation to a batch scheduler. Requests for resources (jobs) are immediately placed into a queue and serviced in order, when resources become available, according to the policies of the scheduler. As a result, jobs often face long, unpredictable queue times, especially jobs that require large numbers of resources or have long runtimes. The allocation of resources and binding of jobs to those resources are fundamentally tied together and out of the user's control. This is unfortunate for workflows because often the overheads of scheduling jobs on these platforms are high, and for a workflow containing many tasks, the penalty is paid many times, which hurts performance [28].

In clouds the process is reversed. Instead of delegating allocation to the resource manager, the user directly provisions the resources required and schedules their computations using a user-controlled scheduler. This provisioning model is ideal for workflows and other loosely-coupled applications because it enables the application to allocate a resource once and use it to execute many tasks, which reduces the total scheduling overhead and can dramatically improve workflow performance [28, 41, 45, 46]. Although in clusters and grids, pilot job systems, such as Condor glide-ins [13], aim to simulate resource provisioning, they face limitations imposed by the target systems, for example, the maximum walltime a job is allowed to run on a resource.

## 4.2.2 On-Demand

Cloud platforms allocate resources on-demand. Cloud users can request, and expect to obtain, sufficient resources for their needs at any time. This feature of clouds has been called the "illusion of infinite resources." The drawback of this approach is that, unlike best-effort queuing, it does not provide an opportunity to wait. If sufficient resources are not available to service a request immediately, then the request fails.

On-demand provisioning allows workflows to be more opportunistic in their choice of resources. Unlike tightly coupled applications, which need all their resources up-front and would prefer to wait in a queue to ensure priority, a workflow application can start with only a portion of the total resources desired. The minimum usable resource pool for workflows containing only serial tasks is one processor. With on-demand provisioning a workflow can allocate as many resources as possible and start making progress immediately.

## 4.2.3 Elasticity

In addition to provisioning resources on-demand, clouds also allow users to return resources on-demand. This dual capability, called elasticity, is a very useful feature for workflow applications because it enables workflow systems to easily grow and shrink the available resource pool as the needs of the workflow change over time. Common workflow structures such as data distribution and data aggregation can significantly change the amount of parallelism in a workflow over time [7]. These changes lead naturally to situations in which it may be profitable to acquire or release resources to more closely match the needs of the application and ensure that resources are being fully utilized.

### 4.2.4 Legacy Applications

Workflow applications frequently consist of a collection of complementary software components developed at different times for different uses by different people. Part of the job of a workflow management system is to weave these heterogeneous components into a single coherent application. Often this must be done without changing the components themselves. In some cases no one wants to modify codes that have been designed and tested many years ago in fear of introducing bugs that may affect the scientific validity of outputs. This can be challenging depending on the environment for which the components were developed and the assumptions made by the developer about the layout of the filesystem. These components are often brittle and require a specific software environment to execute successfully.

Clouds and their use of virtualization technology may make these legacy codes much easier to run. Virtualization enables the environment to be customized to suit the application. Specific operating systems, libraries, and software packages can be installed, directory structures required by the application can be created, input data can be copied into specific locations, and complex configurations can be constructed. The resulting environment can be bundled up as a virtual machine image and redeployed on a cloud to run the workflow.

### 4.2.5 Provenance and Reproducibility

Provenance is the origin and history of an object [8]. In computational science, provenance refers to the storage of metadata about a computation that can be used to answer questions about the origins and derivation of data produced by that computation. As such, provenance is the computational equivalent of a lab scientist's notebook and is a critical component of reproducibility, the cornerstone of experimental science.

Virtualization is a useful feature for provenance because it allows one to capture the exact environment that was used to perform a computation, including all of the software and configuration used in that environment. In previous work [23] we proposed a provenance model for workflow applications in which virtual machine images are a critical component. The idea behind this model is that, if a workflow is executed using a virtual machine, then the VM image can be stored along with the provenance of the workflow. Storing the VM image enables the scientist to answer many important questions about the results of a workflow run such as: What version of the simulation code was used to produce the data? Which library was used? How was the software installed and configured? It also enables the scientist to redeploy the VM image and create exactly the same environment that was used to run the original experiment. This environment could be used to rerun the experiment, or it could be modified to run a new experiment. These capabilities are enabled by the use of virtualization in cloud computing.

## 4.3  Deploying Workflows in the Cloud

Workflow Management Systems, such as Pegasus WMS [16, 40], plan, execute, and monitor scientific workflows. Pegasus WMS, which consists of the Pegasus mapper [14], the DAGMan execution engine [12], and the Condor *schedd* [19] for task execution, performs several critical functions. It adapts workflows to the target execution environment, it manages task execution on distributed resources, it optimizes workflow performance to reduce time to solution and produce scientific results faster, it provides reliability during execution so that scientists need not manage a potentially large number of failures, and they track data so that it can be easily located and accessed during and after workflow execution.

An approach to running workflows on the cloud is to build a virtual cluster using the cloud resources and schedule workflow tasks to that cluster.

### 4.3.1  Virtual Clusters

Scientific workflows require large quantities of compute cycles to process tasks. In the cloud these cycles are provided by virtual machines. To achieve the performance required for large-scale workflows, many virtual machine instances must be used simultaneously. These collections of VMs are called "virtual clusters" [10, 17, 32]. The process of deploying and configuring a virtual cluster is known as contextualization [31]. Contextualization involves complex configuration steps that can be tedious and error-prone to perform manually. To automate this process, software such as the Nimbus Context Broker [31] can be used. This software gathers information about the virtual cluster and uses it to generate configuration files and start services on cluster VMs.

### 4.3.2  Resource Management

Having a collection of virtual machines is not sufficient to run a workflow. Additional software is needed to bind workflow tasks to resources for execution. The easiest way to do this is to use some off-the-shelf resource management system such as Condor [35], PBS [5] or Sun Grid Engine [20]. In this way the virtual cluster mimics a traditional computational cluster. A typical virtual cluster is composed of a virtual machine that acts as a head node to manage the other machines in the cluster and accept tasks from the workflow management system and a set of worker nodes that execute tasks. Configuration of the resource manager and registration of worker nodes with the head node is part of the process of contextualization.

### 4.3.3 Data Storage

Workflows are loosely coupled parallel applications that consist of a set of discrete computational tasks logically connected via data- and control-flow dependencies. Unlike tightly coupled applications in which tasks communicate by sending message via the cluster network, workflow tasks typically communicate using files, where each task produces one or more output files that become input files to other tasks. These files are passed between tasks either through a shared storage system or using some file transfer mechanism.

In order to achieve good performance, these storage systems must scale well to handle data from multiple workflow tasks running in parallel on separate nodes. When running on HPC systems, workflows can usually make use of a high-performance, parallel filesystem such as Lustre [36], GPFS [43], or Panasas [27]. In the cloud, workflows can either make use of a cloud storage service, or they can deploy their own shared filesystem. To use a cloud storage service, the workflow management system would likely need to change the way it manages data. For example, to use Amazon S3 [1] a workflow task needs to fetch input data from S3 to a local disk, perform its computation, then transfer output data from the local disk back to S3. Making multiple copies in this way can reduce workflow performance. Another alternative would be to deploy a filesystem in the cloud that could be used by the workflow. For example, in Amazon EC2 an extra VM can be started to host an NFS filesystem, and worker VMs can mount that filesystem as a local partition. If better performance is needed, then several VMs can be started to host a parallel filesystem such as PVFS [34, 50] or GlusterFS [25].

## 4.4 Case Study: Scientific Workflows on Amazon EC2

To illustrate how clouds can be used for workflow applications, we present a case study using Amazon's EC2 [1] cloud. EC2 is a commercial cloud that exemplifies the IaaS model. It provides computational resources in the form of virtual machine instances, which come in a variety of hardware configurations and are capable of running several different virtualized operating systems. For comparison, we ran workflows on both EC2 and NCSA's Abe cluster [37]. Abe is a typical example of the resources available to scientists on the national cyberinfrastructure. Running the

**Fig. 4.1** The workflow
management in the context of
the execution environment

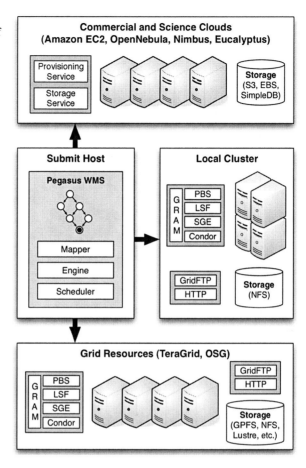

workflows on both platforms allows us to compare the performance, features, and
cost of a typical cloud environment to that of a typical grid environment. Figure 4.1
shows the high-level set up.

## 4.4.1 Applications Tested

Three workflow applications were chosen for this study: an astronomy application
(Montage), a seismology application (Broadband), and a bioinformatics application
(Epigenome). These applications were selected because they cover a wide range of
science domains and resource requirements.

Montage [6] creates science-grade astronomical image mosaics from data collected using telescopes. The size of a Montage workflow (Fig. 4.2) depends upon
the area of the sky (in square degrees) covered by the output mosaic. In our experiments we configured Montage workflows to generate an 8-degree mosaic. The

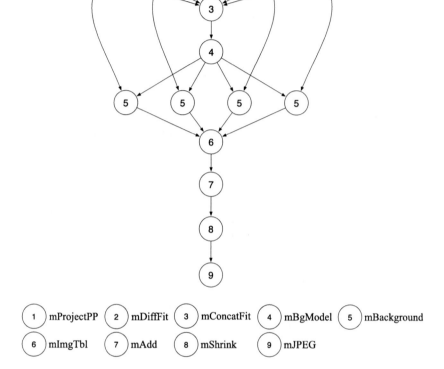

**Fig. 4.2** Montage workflow

resulting workflow contains 10,429 tasks, reads 4.2 GB of input data, and produces 7.9 GB of output data.

Broadband [9] generates and compares seismograms from several high- and low-frequency earthquake simulation codes. Each Broadband workflow (Fig. 4.3) generates seismograms for several sources (scenario earthquakes) and sites (geographic locations). For each (source, site) combination, the workflow runs several high- and low-frequency earthquake simulations and computes intensity measures of the resulting seismograms. In our experiments we used four sources and five sites to generate a workflow containing 320 tasks that reads 6 GB of input data and writes 160 MB of output data.

Epigenome [30] maps short DNA segments collected using high-throughput gene sequencing machines to a previously constructed reference genome using the MAQ software [33]. The workflow (Fig. 4.4) splits several input segment files into small chunks, reformats and converts the chunks, maps the chunks to a reference genome,

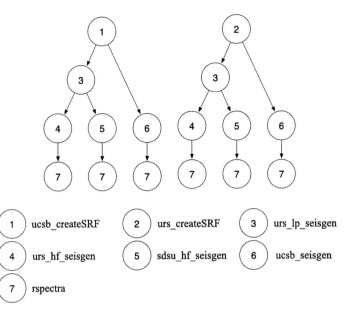

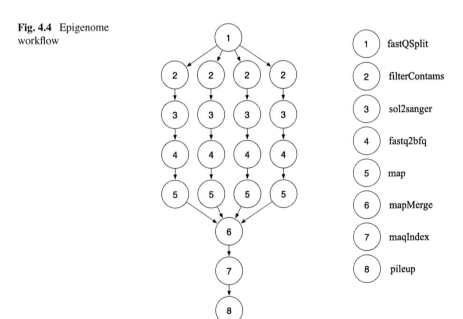

**Fig. 4.3**  Broadband workflow

**Fig. 4.4**  Epigenome workflow

merges the mapped sequences into a single output map, and computes the sequence density for each location of interest in the reference genome. The workflow used in our experiments maps human DNA sequences to a reference chromosome 21. The workflow contains 81 tasks, reads 1.8 GB of input data, and produces 300 MB of output data.

### 4.4.2 Software

All workflows were planned and executed using the Pegasus WMS. Pegasus is used to transform abstract, resource-independent workflow descriptions into concrete, platform-specific execution plans. These plans are executed using DAGMan to track dependencies and release tasks as they become ready, and Condor schedd to run tasks on the available resources.

### 4.4.3 Hardware

EC2 provides resources with various hardware configurations. Table 4.1 compares the resource types used for the experiments. It lists five resource types from EC2 (m1.* and c1.*) and two resource types from Abe (abe.local and abe.lustre). There are several noteworthy details about the resources shown. First, although there is actually only one type of Abe node, there are two types listed in the table: abe.local and abe.lustre. The actual hardware used for these types is equivalent; the difference is in how I/O is handled. The abe.local type uses a local partition for I/O, and the abe.lustre type uses a Lustre partition for I/O. Using two different names is simply a notational convenience. Second, in terms of computational capacity, the c1.xlarge resource type is roughly equivalent to the abe.local resource type with the exception

**Table 4.1** Resource types used

| Type | Arch. | CPU | Cores | Memory | Network | Storage | Price |
|------|-------|-----|-------|--------|---------|---------|-------|
| m1.small | 32-bit | 2.0–2.6 GHz Opteron | 1/2 | 1.7 GB | 1-Gbps Ethernet | Local disk | $0.10/hr |
| m1.large | 64-bit | 2.0–2.6 GHz Opteron | 2 | 7.5 GB | 1-Gbps Ethernet | Local disk | $0.40/hr |
| m1.xlarge | 64-bit | 2.0–2.6 GHz Opteron | 4 | 15 GB | 1-Gbps Ethernet | Local disk | $0.80/hr |
| c1.medium | 32-bit | 2.33–2.66 GHz Xeon | 2 | 1.7 GB | 1-Gbps Ethernet | Local disk | $0.20/hr |
| c1.xlarge | 64-bit | 2.33–2.66 GHz Xeon | 8 | 7.5 GB | 1-Gbps Ethernet | Local disk | $0.80/hr |
| abe.local | 64-bit | 2.33 GHz Xeon | 8 | 8 GB | 10-Gbps InfiniBand | Local disk | N/A |
| abe.lustre | 64-bit | 2.33 GHz Xeon | 8 | 8 GB | 10-Gbps InfiniBand | Lustre | N/A |

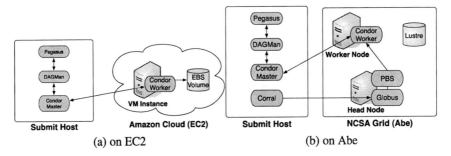

(a) on EC2                              (b) on Abe

**Fig. 4.5** Execution environment

that abe.local has slightly more memory. We use this fact to estimate the virtualization overhead for our test applications on EC2. Third, in rare cases, EC2 assigns Xeon processors for m1.* instances, but for all of the experiments reported here, the m1.* instances used were equipped with Opteron processors. The only significance is that Xeon processors have better floating-point performance than Opteron processors (4 FLOP/cycle vs. 2 FLOP/cycle). Finally, the m1.small instance type is shown having core. This is possible because of virtualization. EC2 nodes are configured to give m1.small instances access to the processor only 50% of the time. This allows a single processor core to be shared equally between two separate m1.small instances.

### 4.4.4 Execution Environment

The execution environment was deployed on EC2 as shown in Fig. 4.5 (left). A submit host running outside the cloud was used to coordinate the workflow, and worker nodes were started inside the cloud to execute workflow tasks. For the Abe experiments, Globus [47] and Corral [11, 29] were used to deploy Condor glideins [22] as shown in Fig. 4.5(right). The glideins started Condor daemons on the Abe worker nodes, which contacted the submit host and were used to execute workflow tasks. This approach creates an execution environment on Abe that is equivalent to the EC2 environment.

### 4.4.5 Storage

Amazon provides several storage services that can be used with EC2. The Simple Storage Service (S3) [39] is an object-based storage service. It supports PUT, GET, and DELETE operations for untyped binary objects up to 5 GB in size. The Elastic Block Store (EBS) [2] is a block-based storage service that provides SAN-like storage volumes up to 1 TB in size. These volumes appear as standard block devices when attached to an EC2 instance and can be formatted with standard UNIX filesystems. EBS volumes cannot be shared between multiple instances.

In comparison, Abe provides shared storage on a large Lustre [36] parallel filesystem. This is a very common storage configuration for cluster and grid platforms.

To run workflow applications storage must be allocated for application executables, input data, intermediate data, and output data. In a typical workflow application, executables are preinstalled on the execution site, input data is copied from an archive to the execution site, and output data is copied from the execution site to an archive.

For all experiments, executables and input data were prestaged to the execution site, and output data were not transferred from the execution site. For the EC2 experiments, executables were installed in the VM images, input data was stored on EBS volumes, and intermediate and output data were written to a local partition. For the Abe experiments, executables and input data were kept on the Lustre filesystem, and intermediate and output data were written in some cases to a local partition (abe.local experiments) and in other cases to the Lustre filesystem (abe.lustre experiments).

EBS was chosen to store input data for a number of reasons. First, storing inputs in the cloud obviates the need to transfer input data repeatedly. This saves both time and money because transfers cost more than storage. Second, using EBS avoids the 10-GB limit on VM images, which is too small to include the input data for all the applications tested. We can access the data as if it were on a local disk without packaging it in the VM image. A simple experiment using the disk copy utility "dd" showed similar performance reading from EBS volumes and the local disk (74.6 MB/s for local, and 74.2 MB/s for EBS). Finally, using EBS simplifies our setup by allowing us to reuse the same volume for multiple experiments. When we need to change instances, we just detach the volume from one instance and reattach it to another.

### 4.4.6 Performance Comparison

In this section we compare the performance of the selected workflow applications by executing them on Abe and EC2. The critical performance metric we are concerned with is the runtime of the workflow (also known as the makespan), which is the total amount of wall clock time from the moment the first workflow task is submitted until the last task completes. The runtimes reported for EC2 do not include the time required to install and boot the VM, which typically averages between 70 and 90 seconds [26]. Now this is much less if you use EBS to store images, and the runtimes reported for Abe do not include the time glidein jobs spend waiting in the queue, which is highly dependent on the current system load. Also, the runtimes do not include the time required to transfer input and output data (see Table 4.4). We assume that this time will be variable depending on WAN conditions. A study of bandwidth to/from Amazon services is presented in [39]. In our experiments we typically observed bandwidth on the order of 500–1000 KB/s between Amazon's East Region datacenter and our submit host in Marina del Rey, CA.

**Fig. 4.6** Runtime
comparison

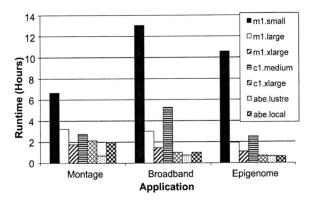

**Table 4.2** Monthly storage
cost

| Application | Volume size | Monthly cost |
|---|---|---|
| Montage | 5 GB | $0.66 |
| Broadband | 5 GB | $0.60 |
| Epigenome | 2 GB | $0.26 |

We estimate the virtualization overhead for each application by comparing the runtime on c1.xlarge with the runtime on abe.local. Measuring the difference in runtime between these two resource types should provide a good estimate of the cost of virtualization.

Figure 4.6 shows the runtime of the selected applications using the resource types shown in Table 4.2. In all cases the m1.small resource type had the worst runtime by a large margin. This is not surprising given its relatively low capabilities.

For Montage, the best EC2 performance was achieved on the m1.xlarge type. This is likely due to the fact that m1.xlarge has twice as much memory as the next best resource type. The extra memory is used by the Linux kernel for the filesystem buffer cache to reduce the amount of time the application spends waiting for I/O. This is particularly beneficial for Montage, which is very I/O-intensive. The best overall performance for Montage was achieved using the abe.lustre configuration, which was more than twice as fast as abe.local. This large gap suggests that having a parallel filesystem is a significant advantage for I/O-bound applications like Montage. The difference in runtime between the c1.xlarge and abe.local experiments suggests that the virtualization overhead for Montage is less than 8%.

The best overall runtime for Broadband was achieved by using the abe.lustre resource type, and the best EC2 runtime was achieved using the c1.xlarge resource type. This is despite the fact that only six of the eight cores on c1.xlarge and abe.lustre could be used due to memory limitations. Unlike Montage, the difference between running Broadband on a relatively slow local disk (abe.local) and running on the parallel filesystem (abe.lustre) is not as significant. This is attributed to the lower I/O requirements of Broadband. Broadband performs the worst on m1.small and c1.medium, which also have the lowest amount memory (1.7 GB). This is because m1.small has only half a core, and c1.medium can only use one of its two cores because of memory limitations. The difference between the runtime using c1.xlarge and the runtime using abe.local was only about 1%. This small difference suggests a relatively low virtualization penalty for Broadband.

For Epigenome, the best EC2 runtime was achieved using c1.xlarge, and the best overall runtime was achieved using abe.lustre. The primary factor affecting the performance of Epigenome was the availability of processor cores, with more cores resulting in a lower runtime. This is expected given that Epigenome is almost entirely CPU-bound. The difference between the abe.lustre and abe.local runtimes was only about 2%, which is consistent with the fact that Epigenome has relatively low I/O and is therefore less affected by the parallel filesystem. The difference between the abe.local and the c1.xlarge runtimes suggests that the virtualization overhead for this application is around 10%, which is higher than both Montage and Broadband. This may suggest that virtualization has a larger impact on CPU-bound applications.

Based on these experiments, we believe that the performance of workflows on EC2 is reasonable given the resources that can be provisioned. Although the EC2 performance was not as good as the performance on Abe, most of the resources provided by EC2 are also less powerful. In the cases where the resources are similar, the performance was found to comparable. The EC2 c1.xlarge type, which is nearly equivalent to abe.local, delivered performance that was nearly the same as abe.local in our experiments.

For I/O-intensive workflows like Montage, EC2 is at a significant disadvantage because of the lack of high-performance parallel filesystems. While such a filesystem could conceivably be constructed from the raw components available in EC2, the cost of deploying such a system would be prohibitive. In addition, because EC2 uses commodity networking equipment, it is unlikely that there would be a significant advantage in shifting I/O from a local partition to a parallel filesystem across the network, because the bottleneck would simply shift from the disk to the network interface. In order to compete performance-wise with Abe for I/O-intensive applications, Amazon would need to deploy both a parallel filesystem and a high-speed interconnect.

For memory-intensive applications like Broadband, EC2 can achieve nearly the same performance as Abe as long as there is more than 1 GB of memory per core. If there is less, then some cores must sit idle to prevent the system from running out of memory or swapping. This is not strictly an EC2 problem, the same issue affects Abe as well.

For CPU-intensive applications like Epigenome, EC2 can deliver comparable performance given equivalent resources. The virtualization overhead does not seem to be a significant barrier to performance for such applications. In fact, the virtualization overhead measured for all application less than 10%. This is consistent with previous studies that show similar virtualization overheads [4, 21, 49]. As such, virtualization does not seem, by itself, to be a significant performance problem for clouds. As virtualization technologies improve, it is likely that what little overhead there is will be further reduced or eliminated.

## 4.4.7  Cost Analysis

In this section we analyze the cost of running workflow applications in the cloud. We consider three different cost categories: resource cost, storage cost, and transfer cost. Resource cost includes charges for the use of VM instances in EC2; storage cost includes charges for keeping VM images in S3 and input data in EBS; and transfer cost includes charges for moving input data, output data, and log files between the submit host and EC2.

**Resource Cost**

Each of the five resource types Amazon offers is charged at a different hourly rate: $0.10/hr for m1.small, $0.40/hr for m1.large, $0.80/hr for m1.xlarge, $0.20/hr for c1.medium, and $0.80/hr for c1.xlarge. Usage is rounded up to the nearest hour, so any partial hours are charged as full hours.

Figure 4.7 shows the per-workflow resource cost for the applications tested. Although it did not perform the best in any of our experiments, the most cost-effective instance type was c1.medium, which had the lowest execution cost for all three applications.

**Storage Cost**

Storage cost consists of (a) the cost to store VM images in S3 and (b) the cost of storing input data in EBS. Both S3 and EBS use fixed monthly charges for the storage of data and variable usage charges for accessing the data. The fixed charges are $0.15 per GB-month for S3 and $0.10 per GB-month for EBS. The variable charges are $0.01 per 1,000 PUT operations and $0.01 per 10,000 GET operations for S3, and $ 0.10 per million I/O operations for EBS. We report the fixed cost per month, and the total variable cost for all experiments performed.

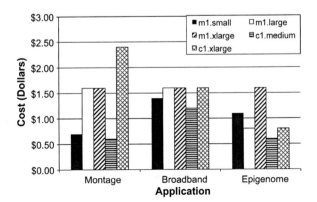

**Fig. 4.7**  Resource cost comparison

We used a 32-bit and a 64-bit VM image for all of the experiments in this paper. The 32-bit image was 773 MB, and the 64-bit image was 729 MB for a total fixed cost of $0.22 per month. In addition, there were 4616 GET operations and 2560 PUT operations for a total variable cost of approximately $0.03.

The fixed monthly cost of storing input data for the three applications is shown in Table 4.2. In addition, there were 3.18 million I/O operations for a total variable cost of $0.30.

**Transfer Cost**

In addition to resource and storage charges, Amazon charges $0.10 per GB for transfer into, and $0.17 per GB for transfer out of, the EC2 cloud. Tables 4.3 and 4.4 show the per-workflow transfer sizes and costs for the three applications studied. Input is the amount of input data to the workflow, output is the amount of output data, and logs is the amount of logging data that is recorded for workflow tasks and transferred back to the submit host. The cost of the protocol used by Condor to communicate between the submit host and the workers is not included, but it is estimated to be less than $0.01 per workflow.

The first thing to consider when provisioning resources on EC2 is the tradeoff between performance and cost. In general, EC2 resources obey the aphorism "you get what you pay for"—resources that cost more perform better than resources that cost less. For the applications tested, c1.medium was the most cost-effective resource type even though it did not have the lowest hourly rate, because the type with the lowest rate (m1.small) performed so badly.

Another important thing to consider when using EC2 is the tradeoff between storage cost and transfer cost. Users have the option of either (a) transferring input data for each workflow separately or (b) transferring input data once, storing it in the cloud, and using the stored data for multiple workflow runs. The choice of which approach to employ will depend on how many times the data will be used, how long the data will be stored, and how frequently the data will be accessed. In general,

**Table 4.3** Per-workflow transfer sizes

| Application | Input | Output | Logs |
|---|---|---|---|
| Montage | 4291 MB | 7970 MB | 40 MB |
| Broadband | 4109 MB | 159 MB | 5.5 MB |
| Epigenome | 1843 MB | 299 MB | 3.3 MB |

**Table 4.4** Per-workflow transfer costs

| Application | Input | Output | Logs | Total |
|---|---|---|---|---|
| Montage | $0.42 | $1.32 | <$0.01 | $1.75 |
| Broadband | $0.40 | $0.03 | <$0.01 | $0.43 |
| Epigenome | $0.18 | $0.05 | <$0.01 | $0.23 |

storage is more cost-effective for input data that is reused often and accessed frequently, and transfer is more cost-effective if data will be used only once. For the applications tested in this paper, the monthly cost to store input data is only slightly more than the cost to transfer it once. Therefore, for these applications, it is usually more cost-effective to store the input data rather than transfer the data for each workflow.

Although the cost of transferring input data can be easily amortized by storing it in the cloud, the cost of transferring output data may be more difficult to reduce. For many applications, the output data is much smaller than the input data, so the cost of transferring it out may not be significant. This is the case for Broadband and Epigenome, for example. For other applications, the large size of output data may be cost-prohibitive. In Montage, for example, the output is actually larger than the input and costs nearly as much to transfer as it does to compute. For these applications, it may be possible to leave the output in the cloud and perform additional analyses there rather than to transfer it back to the submit host.

In [15] the cost of running 1-, 2-, and 4-degree Montage workflows on EC2 was studied via simulation. That paper found the lowest total cost of a 1-degree workflow to be $0.60, a 2-degree to be $2.25, and a 4-degree to be $9.00. In comparison, we found the total cost of an 8-degree workflow, which is 4 times larger than a 4-degree workflow, to be approximately $1.25 if data is stored for an entire month and $2.35 if data is transferred. This difference is primarily due to an underestimate of the performance of EC2 that was used in the simulation, which produced much longer simulated runtimes.

Finally, the total cost of all the experiments presented in this paper was $149.55. That includes all charges related to learning to use EC2, creating VM images, and running test and experimental workflows.

## 4.5  Challenges

In the experiments above we focused on running a workflow application on a single multicore virtual instance. There are several challenges that need to be addressed when running workflows on multiple virtual instances. Here we describe the challenges related to data storage, communications, and configurability.

### 4.5.1  Lack of Appropriate Storage Systems

Existing workflow systems often rely on parallel and distributed filesystems. These are required to ensure that tasks landing on any node can access the outputs of previous tasks that may have executed on another node. It is possible to transfer inputs and outputs for each task separately; however the repeated movement of data is highly inefficient and time-consuming. In addition, it may be costly in a commercial cloud that charges by the number of bytes transferred. Commercial clouds often deploy structured or object-based storage services that can be utilized by workflow applications. However, these services typically do not provide standard filesystem interfaces. In order to use these systems, the application codes must either be modified to interface with the storage services or must be wrapped with additional workflow components that know how to do the translation. Another solution is to deploy a temporary shared filesystem in the cloud as part of a virtual cluster, but the problems with this solution are that it is complex, potentially costly, and requires an additional step to ensure that desired outputs are transferred to permanent storage. A better solution would be a permanent, scalable, parallel filesystem similar to what existing clusters and grids use. The challenge to this approach is that it is not clear how such a filesystem would be provisioned and shared among different users within a cloud. In particular, authentication and authorization would be key challenges of such a system.

### 4.5.2  Relatively Slow Networks

Most communication in workflows occurs through intermediate files that are written by one task and read by a subsequent task. In a distributed environment these files need to be transferred across the network in order for the workflow to make progress. As such, workflows depend on high-performance networks to achieve good performance. This is especially true for data-intensive workflows. Networks that provide high throughput, but not necessarily low latency, are ideal, however the predominant networking technology employed by existing commercial clouds is Gigabit Ethernet. In comparison, many cluster and grid infrastructures provide interconnects that deliver 10 Gigabit/second or better. In order for clouds to be a viable alternative to clusters and grids as a platform for workflow applications, they would need to deploy much faster networks. Making high-performance networking function with OS-level virtualization should be a top priority for future cloud platforms.

### 4.5.3  Lack of Tools

Setting up an environment to run workflows in the cloud is a complex endeavor. There is some work in virtual appliances [42], but those are typically designed for single nodes and not for clusters of nodes. The Nimbus Context Broker [31] can be used to construct virtual clusters and is immensely useful for running workflows in the cloud. More tools are needed to simplify the management of cloud-based environments.

## 4.6  Summary and Future Outlook

The benefit of cloud computing for science is not necessarily in its utility computing and economic aspects, which are not new for academic computing. The benefit of clouds is rather in its technological features that stem from service-oriented architecture and virtualization. In the area of scientific workflows, clouds have many important benefits. These benefits include the illusion of infinite resources, lease-based provisioning, elasticity, support for legacy applications and environments, provenance and reproducibility, and others.

In our work, we supported the workflow execution on the cloud through the deployment of a Condor-based virtual cluster on top of virtual instance. Workflows can also be made to run across multiple virtual instances, but additional communicate of data files need to be supported.

Cloud platforms like Amazon EC2 can be used to execute workflows today, but in the future much work is needed to bring these platforms up to the performance level of the grid. This includes developing cloud storage systems that are appropriate for workflow and other science applications as well as tools to help scientists and workflow engineers deploy their applications in the cloud.

In the future we will see many new cloud-based solutions for workflow applications. For example, we will likely see the development of new management tools that help users run workflows using existing tools in the grid. We may see new workflow systems that are designed with the specific features of the cloud in mind. We will see research on how to deploy workflows across grids and clouds. We may see PaaS and SaaS clouds that are developed exclusively for workflow applications. For example, a PaaS cloud may provide a user-centered Replica Location Service (RLS) for locating files in the cloud and outside, a dynamic Network Attached Storage (NAS) service for storing files used and created by workflows, a transformation catalog service to store and manage application binaries, and a workflow execution service for managing tasks and dependencies. A SaaS cloud for workflows may provide an application-specific portal where a user could enter the details of a desired computation and have the underlying workflow services generate a new workflow instance, plan and execute the computation, and provide access to the results.

**Acknowledgements**   We acknowledge the contributions of Karan Vahi, Gaurang Mehta, Phil Maechling, Benjamin P. Berman, and Bruce Berriman. This work was supported by the National Science Foundation under the SciFlow (CCF-0725332) grant. This research made use of Montage, funded by the National Aeronautics and Space Administration's Earth Science Technology Office, Computation Technologies Project, under Cooperative Agreement Number NCC5-626 between NASA and the California Institute of Technology.

# References

1. Amazon.com: Amazon web services (aws). http://aws.amazon.com
2. Amazon.com: Elastic block store (ebs). http://aws.amazon.com/ebs
3. Armbrust, M., Fox, A., Griffith, R., Joseph, A.D., Katz, R., Konwinski, A., Lee, G., Patterson, D., Rabkin, A., Stoica, I., Zaharia, M.: Above the clouds: a Berkeley view of cloud computing. Tech. rep., UC Berkeley (2009)
4. Barham, P., Dragovic, B., Fraser, K., Hand, S., Harris, T., Ho, A., Neugebauer, R., Pratt, I., Warfield, A.: Xen and the art of virtualization. In: Proceedings of the 19th ACM Symposium on Operating Systems Principles (2003)

5. Bayucan, A., Henderson, R.L., Lesiak, C., Mann, B., Proett, T., Tweten, D.: Portable batch system: external reference specification. Tech. rep., MRJ Technology Solutions (1999)

6. Berriman, B., Bergou, A., Deelman, E., Good, J., Jacob, J., Katz, D., Kesselman, C., Laity, A., Singh, G., Su, M.H., Williams, R.: Montage: a grid-enabled image mosaic service for the NVO. In: Astronomical Data Analysis Software and Systems (ADASS) XIII (2003)

7. Bharathi, S., Chervenak, A., Deelman, E., Mehta, G., Su, M.H., Vahi, K.: Characterization of scientific workflows. In: Proceedings of the 3rd Workshop on Workflows in Support of Large-Scale Science (WORKS'08) (2008)

8. Bruneman, P., Khanna, S., Tan, W.C.: Why and where: a characterization of data provenance. In: Proceedings of the 8th International Conference on Database Theory (2001)

9. Center, S.C.E.: Community modeling environment. http://www.scec.org/cme/

10. Chase, J.S., Irwin, D.E., Grit, L.E., Moore, J.D., Sprenkle, S.E.: Dynamic virtual clusters in a grid site manager. In: 12th IEEE International Symposium on High Performance Distributed Computing (HPDC'03) (2003)

11. Corral. http://pegasus.isi.edu/corral/latest

12. Dagman (directed acyclic graph manager). http://cs.wisc.edu/condor/dagman

13. Deelman, E., Gannon, D., Shields, M., Taylor, I.: Workflows and e-Science: an overview of workflow system features and capabilities. Future Gener. Comput. Syst. **25**(5), 528–540 (2008)

14. Deelman, E., Livny, M., Mehta, G., Pavlo, A., Singh, G., Su, M.H., Vahi, K., Wenger, R.K.: Pegasus and DAGMan from Concept to Execution: Mapping Scientific Workflows Onto Today's Cyberinfrastructure, pp. 56–74. IOS, Amsterdam (2008)

15. Deelman, E., Singh, G., Livny, M., Berriman, B., Good, J.: The cost of doing science on the cloud: the montage example. In: Proceedings of the 2008 ACM/IEEE Conference on Supercomputing (2008)

16. Deelman, E., Singh, G., Su, M.H., Blythe, J., Gil, Y., Kesselman, C., Mehta, G., Vahi, K., Berriman, G.B., Good, J., Laity, A., Jacob, J.C., Katz, D.S.: Pegasus: a framework for mapping complex scientific workflows onto distributed systems. Sci. Program. **13**(3), 219–237 (2005)

17. Foster, I., Freeman, T., Keahey, K., Scheftner, D., Sotomayor, B., Zhang, X.: Virtual clusters for grid communities. In: Proceedings of the 6th IEEE International Symposium on Cluster Computing and the Grid (CCGRID'06) (2006)

18. Foster, I., Kesselman, C., Tuecke, S.: The anatomy of the grid: enabling scalable virtual organizations. Int. J. High Perform. Comput. Appl. **15**(3), 200–222 (2001)

19. Frey, J., Tannenbaum, T., Foster, I., Livny, M., Tuecke, S.: Condor-G: a computation management agent for multi-institutional grids. In: 10th International Symposium on High Performance Distributed Computing (2001)

20. Gentzsch, W.: Sun grid engine: towards creating a compute power grid. In: Proceedings of the 1st International Symposium on Cluster Computing and the Grid (2001)

21. Gilbert, L., Tseng, J., Newman, R., Iqbal, S., Pepper, R., Celebioglu, O., Hsieh, J., Cobban, M.: Performance implications of virtualization and hyper-threading on high energy physics applications in a grid environment. In: Proceedings of the 19th IEEE International Parallel and Distributed Processing Symposium (IPDPS'05) (2005)

22. Glidein. http://www.cs.wisc.edu/condor/glidein

23. Groth, P., Deelman, E., Juve, G., Mehta, G., Berriman, B.: Pipeline-centric provenance model. In: Proceedings of the 4th Workshop on Workflows in Support of Large-Scale Science (WORKS'09) (2009)

24. Hoffa, C., Mehta, G., Freeman, T., Deelman, E., Keahey, K., Berriman, B., Good, J.: On the use of cloud computing for scientific workflows. In: Proceedings of the 3rd International Workshop on Scientific Workflows and Business Workflow Standards in e-Science (SWBES'08) (2008)

25. Inc., G.: Glusterfs. http://www.gluster.org

26. Inc., H.: Cloudstatus. http://www.cloudstatus.com

27. Inc., P.: Panasas. http://www.panasas.com

28. Juve, G., Deelman, E.: Resource provisioning options for large-scale scientific workflows. In: Proceedings of the 3rd International Workshop on Scientific Workflows and Business Workflow Standards in e-Science (SWBES'08) (2008)
29. Juve, G., Deelman, E., Vahi, K., Mehta, G.: Experiences with resource provisioning for scientific workflows using Corral. Sci. Program. **18**(2), 77–92 (2010)
30. Juve, G., Deelman, E., Vahi, K., Mehta, G., Berriman, B., Berman, B.P., Maechling, P.: Scientific workflow applications on Amazon EC2. In: Workshop on Cloud-based Services and Applications in Conjunction with 5th IEEE International Conference on e-Science (e-Science'09) (2009)
31. Keahey, K., Freeman, T.: Contextualization: providing one-click virtual clusters. In: Proceedings of the 4th International Conference on eScience (eScience'08) (2008)
32. Kee, Y., Kesselman, C., Nurmi, D., Wolski, R.: Enabling personal clusters on demand for batch resources using commodity software. In: Proceedings of the IEEE International Symposium on Parallel and Distributed Processing (IPDPS'08) (2008)
33. Li, H., Ruan, J., Durbin, R.: Mapping short DNA sequencing reads and calling variants using mapping quality scores. Genome Res. **18**(11), 1851–1858 (2008)
34. Ligon, W.B., Ross, R.B.: Implementation and performance of a parallel file system for high performance distributed applications. In: Proceedings of the Fifth IEEE International Symposium on High Performance Distributed Computing (1996)
35. Litzkow, M., Livny, M., Mutka, M.: Condor—a hunter of idle workstations. In: Proceedings of the 8th International Conference on Distributed Computing Systems (1988)
36. Microsystems, S.: Lustre. http://www.lustre.org
37. National center for supercomputing applications (ncsa). http://www.ncsa.illinois.edu
38. Open science grid. http://www.opensciencegrid.org
39. Palankar, M.R., Iamnitchi, A., Ripeanu, M., Garfinkel, S.: Amazon S3 for science grids: a viable solution? In: International Workshop on Data-Aware Distributed Computing (2008)
40. Pegasus workflow management system. http://pegasus.isi.edu
41. Raicu, I., Zhao, Y., Dumitrescu, C., Foster, I., Wilde, M.: Falkon: a fast and light-weight task execution framework. In: Proceedings of the 2007 ACM/IEEE Conference on Supercomputing (2007)
42. Sapuntzakis, C., Brumley, D., Chandra, R., Zeldovich, N., Chow, J., Lam, M., Rosenblum, M.: Virtual appliances for deploying and maintaining software. In: Proceedings of the 17th USENIX Conference on System Administration (2003)
43. Schmuck, F., Haskin, R.: GPFS: a shared-disk file system for large computing clusters. In: Proceedings of the 1st USENIX Conference on File and Storage Technologies (2002)
44. San Diego Supercomputing Center (sdsc). http://www.sdsc.edu
45. Singh, G., Kesselman, C., Deelman, E.: Performance impact of resource provisioning on workflows. Tech. rep., University of Southern California, Information Sciences Institute (2005)
46. Singh, G., Kesselman, C., Deelman, E.: A provisioning model and its comparison with best-effort for performance-cost optimization in grids. In: Proceedings of the 16th International Symposium on High Performance Distributed Computing (HPDC'07) (2007)
47. Sotomayor, B., Childers, L.: Globus Toolkit 4 Programming Java Services. Elsevier/Morgan Kaufmann, Amsterdam (2006)
48. Teragrid. http://www.teragrid.org/
49. Youseff, L., Seymour, K., You, H., Dongarra, J., Wolski, R.: The impact of paravirtualized memory hierarchy on linear algebra computational kernels and software. In: Proceedings of the 17th International Symposium on High Performance Distributed Computing (2008)
50. Yu, W., Vetter, J.S.: Xen-based HPC: a parallel I/O perspective. In: Proceedings of the 8th IEEE International Symposium on Cluster Computing and the Grid (CCGrid'08) (2008)

# Chapter 5
# Auspice: Automatic Service Planning in Cloud/Grid Environments

**David Chiu and Gagan Agrawal**

**Abstract** Recent scientific advances have fostered a mounting number of services and data sets available for utilization. These resources, though scattered across disparate locations, are often loosely coupled both semantically and operationally. This loosely coupled relationship implies the possibility of linking together operations and data sets to answer queries. This task, generally known as automatic service composition, therefore abstracts the process of complex scientific workflow planning from the user. We have been exploring a metadata-driven approach toward automatic service workflow composition, among other enabling mechanisms, in our system, Auspice: Automatic Service Planning in Cloud/Grid Environments. In this paper, we present a complete overview of our system's unique features and outlooks for future deployment as the Cloud computing paradigm becomes increasingly eminent in enabling scientific computing.

## 5.1 Introduction

Steady progress in various scientific fields including, but not limited to, geoinformatics [22, 29, 30], astronomy [41], bioinformatics [26–28], and high-energy physics [5], has led to a cornucopia of new data. These scientific data sets are typically stored in structured, low-level files for a variety of reasons despite the ongoing success in database technologies.[1] To store these vast collections of files, Mass

---

[1]For instance, a series of scientific observations is easily represented by arrays but not relational tables.

---

D. Chiu (✉)
School of Engineering and Computer Science, Washington State University, Vancouver, WA 98686, USA
e-mail: david.chiu@wsu.edu

G. Agrawal
Department of Computer Science and Engineering, The Ohio State University, Columbus, OH 43210, USA
e-mail: agrawal@cse.ohio-state.edu

M. Cafaro, G. Aloisio (eds.), *Grids, Clouds and Virtualization*,
Computer Communications and Networks,
DOI 10.1007/978-0-85729-049-6_5, © Springer-Verlag London Limited 2011

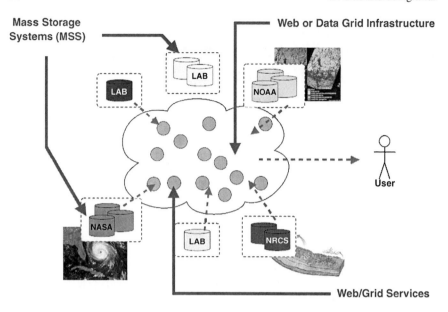

**Fig. 5.1** Scientific data repositories and services

Storage Systems (MSS) are typically employed. Reliant on voluminous, but slow, disk/tape storage, MSS systems are distributed over existing networks including clusters, scientific Data Grids, and more recently, the Cloud [44], in combination that culminates a "scientific Web." Figure 5.1 exemplifies this distribution model of scientific data storage and services.

At the same time, XML [45] emerged, a declarative markup language that allows common users to describe data sets. Having a accessible means for anyone to invent and provide data descriptions, metadata were eventually substantiated and standardized across many domains. As a result, metadata have become essential to many of today's applications. Because scientific data is often cryptic, the data management and Web communities have pressed for the coupling of metadata with certain data sets. The Dublin Core Metadata Initiative, for instance, has instituted a general set of cross-domain metadata elements (e.g., author, title, date, etc.) [18]. Attuned the importance of metadata, the scientific community also began undertaking tremendous efforts toward standardizing metadata formats. These efforts produced such formats as the Content Standard for Digital Geospatial Metadata (CSGDM) [21] and the Biological Data Profile (BDP) [39]. With a mounting number of reliable scientific metadata, relevant data sets can be identified and accessed more efficiently in today's cyberinfrastructures.

Meanwhile, the success of Service-Oriented Architectures (SOA) has ushered an abundant deployment of interoperable processes/services, access to high-level utilities, and other compute and storage resources across scientific and geographic domains. These concrete advancements were instrumental in resurfacing the metacomputing framework, now known as the Grid [23], which is aptly named to reflect the vision of ubiquitous access to pervasive compute resources.

With the Grid's high availability of distributed data sets and services comes the nontrivial challenge for scientists and other end-users to manage such information. For instance, certain information involves execution of several operations, with disparate inputs, in a particular sequence. This process is typically known as service composition. The Grid computing environment, together with the need to manage and process scientific data, triggered a resurgence of interest in workflows, which were originally employed for managing complex business operations. Whereas today's mainstream scientific workflow systems (e.g., Pegasus [19], Kepler [2], Triana [38], Taverna [40], and others) have been instrumental in reducing the need for domain scientists to be familiar with the nuances of large-scale computing. For instance, most scientific workflow management systems provide high-level user interfaces for planning dependent operations, whereas computing provisions such as scheduling and resource allocation are hidden. Certainly, a goal for enabling these service level workflows is to automate their composition while simultaneously hiding such esoteric details as service and data discovery, integration, and scheduling from the common user.

### 5.1.1 Our Vision with Auspice

In this chapter, we discuss our contributions with Auspice (Automatic Service Planning in Cloud/Grid Environments), a metadata-driven service composition system developed at the Ohio State University. Auspice processes queries through automatic composition and execution of service workflow plans. Auspice is data-driven in the way that it leverages domain specific metadata to automatically derive loosely coupled service plans that are semantically sound. Our system also enables Quality of Service through adaptive service planning, allowing it to scale execution times and data derivation accuracies to user needs. Auspice also exploits the emergence of the pay-as-you-go supercomputing infrastructure provisioned via the Cloud, by offering a flexible intermediate data cache. Our system, shown in Fig. 5.2, comprises several independent layers, each encapsulated with certain functionalities. We outline the system in its entirety and, in the same effort, our research contributions.

The rest of this chapter is organized as follows. We first describe the Auspice metadata framework (Sect. 5.2) because it is central to our framework. Methods for planning, QoS handling, and caching are described in Sect. 5.3. Our approach for keyword querying integration is discussed in Sect. 5.4. A system evaluation was performed to show the effectiveness of our QoS handling and caching, two of Auspice's distinguishing features (Sect. 5.5). A discussion of related works is given in Sect. 5.6, and finally, we conclude in Sect. 5.7.

## 5.2 Metadata Framework

Auspice's Semantics Layer in Fig. 5.2 facilitates the metadata functionalities necessary for our automatic service composition algorithm.

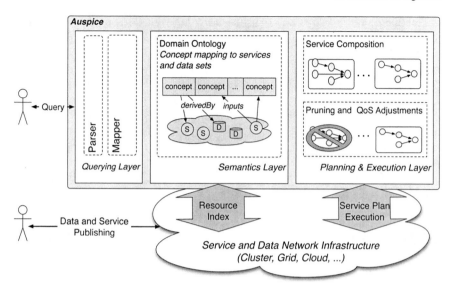

**Fig. 5.2** Auspice system architecture

## 5.2.1 Capturing Concept Derivation

In designing the framework, our observation is that many scientific fields typically contain a set of domain-specific concepts, e.g., in geoinformatics: coordinates, bathymetry (water depth), coast lines, etc. These domain concepts can typically be derived by the available data sets and services. Users, who have become increasingly goal-oriented, target these concepts in their queries: an engineer might prefer a system capable of immediately answering

```
''what is the coast line along coordinate x?''
```

rather than having to find and orchestrate the execution of a workflow composing several operations using files from several potentially disparate data sources. Abstractly, one can envision a "concept derivation" scheme as a means to automatically generate plans, composing the necessary service operations and data sets to answer queries. That is, a concept $c$ is derived by a service $s$, whose inputs $(x, y, z)$ are again substantiated by concepts $c_x, c_y, c_z$, etc., until a terminal element (either a service without input or data file) has been reached.

We proposed an ontology to capture these concept derivation relationships: Let ontology $O = (V_O, E_O)$ be a directed acyclic graph where its set of vertices, $V_O$, comprises a disjoint set of classes: concepts $C$, services $S$, and data types $D$, i.e., $V_O = (C \cup S \cup D)$. Each directed edge $(u, v) \in E_O$ must denote one of the following relations:

- $(u \delta^{c \to^s} v)$: concept-service derivation. Service $u \in S$ used to derive $v \in C$.
- $(u \delta^{c \to^d} v)$: concept-data type derivation. Data type $u \in D$ used to derive $v \in C$.

**Fig. 5.3** Coast line workflow example

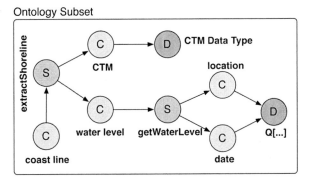

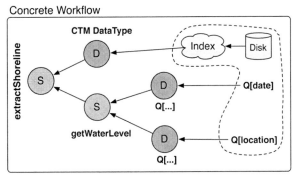

- $(u\delta^{s \rightarrow c}v)$: service-concept derivation. Concept $u \in C$ used to derive service $v \in S$.

Planning a workflow with this structure in place becomes possible as the task is simplified to depth-first search from the query's targeted concept node. In the top half of Fig. 5.3, we show one possible subset of the ontology vis-à-vis with the coast line target concept.[2] Although the abstract workflow is not shown, it can be easily envisioned by reversing the arrows (depicting derivation dependency) and removing the intermediate concept nodes. The bottom half of Fig. 5.3 shows the executable, concrete workflow after some substantiation of the required data (and query parameters).

### 5.2.2 Enabling Fast Resource Identification

Although the aforementioned semantics framework can be used to find a concept's overall derivation structure, inputs and files specific to the user's query has yet to

---

[2]Other derivation paths may exist within a certain ontology, but for simplicity, we show just one here.

be determined. This is by far the most time-consuming aspect of the planning algorithm. But to reduce its complexity, we apply a metadata indexing scheme for data sets and services. In a process we call *metadata registration*, each known file is indexed by their respective popular identifying concepts extracted from their corresponding metadata. For instance, a file in the geographical domain is typically identified by the time and location that it represents. The general metadata registration framework allows domain scientists to define rules for the concept's extraction from each metadata format and builds a unified index to quickly identify the necessary files for workflow construction. Service metadata registration, although less crucial since $|D| \gg |S|$, is also employed to index their derivation and input needs.

On receiving the above coast line query, we have shown that Auspice can accurately plan the workflow toward its derivation (given a set of known services and data sets), as well as quickly identifying the necessary files associated with coordinate $x$. A nuanced discussion of this example within the semantics framework can also be found in [11].

## 5.3 Service Workflow Planning

In this section, we begin the discussion of the system's Planning and Execution Layer, responsible for automatic service composition, making QoS decisions, and workflow execution. The ontology described in the previous section is capable of defining workflows in the recursive form of composite derivations. In the same way, a workflow $w$ can be defined in our system as follows:

$$w = \begin{cases} \varepsilon, \\ d \in D, \\ (op, P_{op}) \in S \end{cases} \tag{1}$$

such that terminals $\varepsilon$ and $d \in D$ denote a null workflow and a data instance (file, user input, intermediate data, etc.) belonging to a specific data type in $D$, respectively. Nonterminal $(op, P_{op}) \in S$ is a tuple where $op$ denotes a service operation with a corresponding parameter list $P_{op} = (p_1, \ldots, p_k)$ and each $p_i$ is itself a workflow. In other words, a workflow is a tuple that either contains a single data instance or a service operation whose parameters are, recursively, (sub)workflows.

Our planning algorithm, called workflow enumeration (WFEnum), takes as input the targeted concept, $t$ and $Q[\ldots]$, the mapped list of query parameters such that $Q[k] \rightarrow val|k \in C$. The planner's goal is to enumerate all possible service plans that can be used to somehow derive $t$ with respect to the given $Q$. Algorithm 1 shows a condensed version of the WFEnum algorithm originally proposed in [11]. While the overall algorithm has been reduced here, its basic logic is still conveyed. Faithful to the workflow structure defined in (1), WFEnum generates and returns a list of possible workflows, $W = (w_1, \ldots, w_m)$, by first deriving $t$ via any data sets available (Lines 2–8). Finally, WFEnum searches all service-derived paths (Lines 9–18)

---

**Algorithm 1** WFEnum($t$, $Q$)

---

1:  $W \leftarrow ()$
2:  **for all** $\{(t, v) \in G | v \in D\}$ **do**
3:      $//v = d \in D$
4:      $F \leftarrow \sigma_{<Q>}(d)$ //select files w.r.t. $Q$
5:      **for all** $f \in F$ **do**
6:          $W \leftarrow W \cup \{f\}$
7:      **end for**
8:  **end for**
9:  **for all** $\{(t, v) \in G | v \in S\}$ **do**
10:     $//v = (op, P_{op}) \in S$
11:     $W_{op} \leftarrow ()$
12:     **for all** $p \in P_{op}$ **do**
13:         $W_{op} \leftarrow W_{op} \times \text{WFEnum}(p.t, Q)$
14:     **end for**
15:     **for all** $pm \in W_{op}$ **do**
16:         $W \leftarrow W \cup \{(op, pm)\}$
17:     **end for**
18: **end for**
19: return $W$

---

by composing all services known to derive $t$. The composition of the services is driven by a recursive reduction of the concepts pertaining to all of its input parameters. Whereas the ontology guides the derivation links (Line 2 for data types and Line 9 for services), the metadata index, implicit in the algorithm, identifies the files efficiently (Line 4).

### 5.3.1 Planning with QoS Adaptivity

Often, there exists multiple ways to derive a given query, using different combinations of data sources and services. Which costs differentiate each workflow, then, should be determined. In most scientific domains, we are concerned with two Qualities of Service constraints: execution time and the accuracy of results.

In terms of time, it is expected that users may have certain constraints on execution time. But in heterogeneous networks, such as geographically diverse Grids, execution times cannot always be guaranteed. In our earlier work [15], we showed that Auspice can adapt to heterogeneity in underlying networks with varying bandwidths by dynamically reducing the complexity of data sets in the workflow. Reducing data sets, however, introduced a vexing problem of finding the corresponding errors with respect to the scientific domain. We return to the running coast line query example. Let us assume that the user now requests that the coast line be returned in some $T$ amount of time, with a mean error no greater than $E$ meters in length. Now, consider a case where Auspice determined a workflow can be completed in $T$ time by reducing the resolution of some image by $\alpha\%$. The problem becomes finding the relationship between the system accuracy parameter, $\alpha$, and the domain-specific error, $E$, for example, difference in meters.

We studied this effect in [14, 16] and proposed a general approach to automatically assign system-based adaptive parameters, such as an $\alpha$ resolution rate, in order to still meet the user-based accuracy/error parameters, $E$. We found that many scientific error models are complex and implemented by read-only algorithms. Clearly, given some scientific error model $\sigma(\alpha)$ that estimates $E$ on varying resolution values it is not always possible to inverse $\sigma$ for providing a precise $\alpha$. We proposed that, given an error constraint $E$, $\sigma$ can be iteratively invoked on disparate values $(v_1, \ldots, v_i)$ until convergence, i.e., $E \approx \sigma(v_i)$, and trivially, $\alpha \approx v_i$. We showed, in [16], that $v_i$ can be found via a binary-search approach, and $\alpha$-convergence is on the order of microseconds, which is negligible when compared to the overall planning time.

### 5.3.2 Flexible Derived Data Caching

As scientific data sets continue to mount, the overall execution times of large-scale workflows clearly become dominated by computation and network transfer times. In an effort to reduce execution times, we believe that a traditional approach, caching, can be applied to the Auspice framework. In [13], we built a hierarchical cache for intermediate derived workflow data. This cache uses a modified version of $B^x$-Trees [33], to index already derived data. Our cache was implemented on a cluster and was shown to not only scale well to an increasing number of cluster nodes, but also speed up our queries accordingly.

To bring this idea a step beyond clusters and into today's Cloud infrastructure [4], we surmise that technologies developed for Web caching can be employed to build a flexible hierarchical cache. A Cloud-based cache will be flexible in the sense that not only can it dynamically allocate Cloud nodes to grow in size, but it should also shrink to save cost for Cloud usage. We utilize consistent hashing [34], which has found application in Web proxies, among others. With all forms of hashing, the dynamic allocation (and deallocation) of nodes causes overhead due to the migration of indexes to new nodes. Consistent hashing was developed to reduce this overhead and hence is friendly to dynamic underlying infrastructures such as the Cloud. For Cloud-use, we proposed an algorithm, Greedy Bucket Allocation (GBA), which allocates Cloud nodes as a last resort to save costs while managing to cache as many intermediate data sets as possible [12]. We have shown that on-demand machine allocation for intermediate data caching within a simulated Cloud not only improves execution times, but also minimizes the potentials for underutilizing resources (not incidentally, it also helps reduce the Cloud utilization cost). While our algorithms were able to scale up resources when needed, the Cloud computing paradigm ushers in a new dimension in cost optimization. That is, applications should also scale down to save cost. However, this decision is difficult to make—data and job migration costs are high, so down-scaling should only be performed when it is predictable that potentials for increasing workloads are not imminent.

We implement a cache contraction scheme to merge nodes when query intensities are lowered. Our scheme is based on a combination of exponential decay and a temporal sliding window. Because the size of our cache system (number of nodes) is highly dependent on the frequency of queries during some timespan, we propose a global cache eviction scheme that captures querying behavior. In our contraction scheme, we employ a streaming model, where incoming query requests represent streaming data, and a global view of the most recently queried keys is maintained in a sliding window. A sliding window, $T = (t_1, \ldots, t_m)$, comprises $m$ time slices of some fixed real-time length. Each time slice $t_i$ associates a set of keys queried in the duration of that slice. We argue that, as time passes, older unreferenced keys (i.e., $t_i$ nearing $t_m$) should have a lower probability of existing in the cache. As these less relevant keys become evicted, the system makes room for newer, incoming keys (i.e., $t_i$ nearing $t_1$) and thus capturing temporal locality of the queries.

Cache eviction occurs when a time slice has reached $t_{m+1}$, and at this time, an eviction score,

$$\lambda(k) = \sum_{i=1}^{m} \alpha^{i-1} \big| \{k \in t_i\} \big|,$$

is computed for every key, $k$, within the expired slice. The ratio $\alpha : 0 < \alpha < 1$ is a *decay* factor, and $|\{k \in t_i\}|$ returns the number of times $k$ appears in some slice $t_i$. Here, $\alpha$ is passive in the sense that a higher value corresponds to a larger amount of keys that is kept in the system. After $\lambda$ has been computed for each key in $t_{m+1}$, any key whose $\lambda$ falls below the threshold, $T_\lambda$, is evicted from the system. Notice that $\alpha$ is amortized in the older time slices, in other words, recent queries for $k$ are rewarded, so $k$ is less likely to be evicted. Clearly, the sliding window eviction method is sensitive to the values of $\alpha$ and $m$. A baseline value for $T_\lambda$ would be $\alpha^{m-1}$, which will not allow the system to evict any key if it was queried even just once in the span of the sliding window. We will show their effects in the experimental section.

Due to the eviction strategy, a set of cache nodes may eventually become lightly loaded, which is an opportunity to scale our system down. The nodes' indices can be merged, and subsequently, the superfluous node instances can be discarded.

## 5.4 Keyword Querying

The Querying Layer in Auspice is responsible for decomposing a user's query into concepts within the ontology, as well as materializing the concepts with the user's given values. Keyword search, without saying, has become a mainstay for common querying. Albeit that abundant effort has been put into supporting keyword searches in general unstructured documents, e.g., the Web, the current technologies are quite excessive in the context of our system. Auspice encompasses domain-specific information, a quality generally missing from the Web. Auspice's knowledge framework, including metadata and the ontology (discussed in the Semantics Layer), can be employed here to better facilitate keyword queries. We believe that this is the first keyword-search endeavor into automatic service composition and scientific workflow systems.

### 5.4.1 Keyword-Maximization Query Planning

To support keyword queries, we automatically compose all workflows relevant to the most number of keywords in the user query, $K$. We currently support only AND-style keyword queries. Auspice's querying algorithm returns all workflow plans, $w$, whose concept-derivation graph, $\psi(w)$ (to be discussed later), contains the most concepts from $K$, while under the constraints of the user's query parameters, $Q$. To exemplify the algorithms, we prescribe the ontology subset shown in Fig. 5.4 to our discussion. Furthermore, we interweave the description of the algorithms with the keyword query example:

``wind coast line CTM image (41.48335,-82.687778) 8/20/2009''

Here, we note that the given coordinates point to Sandusky, Ohio, a location where we have abundant data sets.

The data and service metadata registration procedure, discussed previously, allows the user to supply some keywords that describe their data set or the output of the service. These supplied keywords are used to identify the concepts in which the new resource derives, and if such a concept does not exist, the user is given an option to create one in the ontology. As such, each concept $c$ has an associated set of keywords, $K_c$. For instance, the concept of *elevation* might associate $K_{\text{elevation}} =$

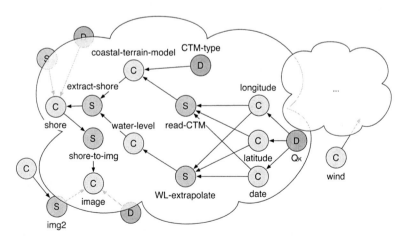

**Fig. 5.4** An exemplifying ontology

{"height", "elevation", "DEM"}. The WordNet database [20] was also employed to expand $K_c$ for the inclusion of each term's synonyms.

Some terms, however, can only be matched by their patterns. For example, "13:00" should be mapped to the concept *time*. Others require further processing. A coordinate $(x, y)$ is first parsed and assigned concepts independently, (i.e., $x \leftarrow$ longitude and $y \leftarrow$ latitude). Because Auspice is currently implemented over the geospatial domain, only a limited number of patterns are expected. Finally, the last pattern involves value assignment. In our keyword system, values can be given directly to concepts using a *keyword* = *value* string. That is, the keyword query "water level $(x, y)$" is equivalent to "water level latitude = $y$ longitude = $x$". Finally, each query term is matched against this set of terms to identify their corresponding concepts. Indeed, a keyword may correspond with more than one concept. However, to be discussed next, using the concept-derivation of the keyword concepts, our algorithm prunes all unlikely terms during the workflow planning phase.

Before we describe the workflow enumeration algorithm, WFEnum_key (shown as Algorithm 2), we introduce the notion of concept derivation graphs (or $\psi$-graphs) which is instrumental in WFEnum_key for pruning. $\psi$-graphs are obtained as concept-derivation relationships, $\psi(c) = (V_\psi, E_\psi)$, where $c$ is a concept, from the ontology. All vertices within $\psi(c)$ denote only concepts, and its edges represent derivation paths. As an aside, $\psi$ can also be applied on workflows, i.e., $\psi(w)$ extracts the concept-derivation paths from the services and data sets involved in $w$.

## 5.4.2 *Planning Algorithm*

WFEnum_key's inputs include $c_t$, which denotes the targeted concept. That is, all generated workflows $w$ must have a $\psi$-graph rooted in concept $c_t$. Specifically, only workflows $w$ whose $\psi(w) \subseteq \psi(c_t)$ will be considered for the result set. The next input, $\Phi$, is a set of required concepts, and every concept in $\Phi$ must be included in the derivation graph of $c_t$. A set of query parameters, $Q$, is also given to this algorithm. These would include the coordinates and the date given by the user in our example query. $Q$ is used to identify the correct files and also as input into services that require these particular concept values. Finally, the ontology, $O$, supplies the algorithm with the derivation graph.

On Lines 2–8, the planning algorithm first considers all data-type derivation possibilities within the ontology for $c_t$, e.g., $(c_t \delta^{c \rightarrow d} d_t)$. All data files are retrieved with respect to data type $d_t$ and the parameters given in $Q$. Each returned file record, $f$, is an independent file-based workflow deriving $t$. Next, the algorithm handles service-based derivations. From the ontology $O$, all $(c_t \delta^{c \rightarrow s} s_t)$ relations are retrieved. Then for each service $s_t$ that derives $c_t$, its parameters must first be recursively planned. Line 15 thus retrieves all concept derivation edges $(s_t \delta^{s \rightarrow c} c_{s_t})$ for each of its parameters. Opportunities for pruning are abundant here. For instance, if the required set of concepts, $\Phi$, is not included in the $\psi$-graphs of all $s_t$'s parameters combined, then $s_t$ can be pruned because it does not meet the query's requirements. For example, on the bottom left corner of Fig. 5.4, we can imply that another service,

---

**Algorithm 2** WFEnum_key($c_t$, $\Phi$, $Q$, $O$)

---

1: static $W$
2: **for all** concept-data derivation edges w.r.t. $c_t$, $(c_t \delta^{c \to d} d_t) \in E_O$ **do**
3:     # data type $d_t$ derives $c_t$; build on $d_t$
4:     $F \leftarrow \sigma_{<Q>}(d_t)$ //select files w.r.t. $Q$
5:     **for all** $f \in F$ **do**
6:         $W \leftarrow W \cup \{f\}$
7:     **end for**
8: **end for**
9: # any workflow enumerated must be reachable within $\Phi$
10: **for all** concept-service derivation edges w.r.t. $c_t$, $(c_t \delta^{c \to s} s_t) \in E_O$ **do**
11:     # service $s_t$ derives $c_t$; build on $s_t$
12:     $W_{s_t} \leftarrow ()$
13:     # remove target, $c_t$, from requirement set (since we current see it)
14:     $\Phi \leftarrow \{\Phi \setminus c_t\}$
15:     **for all** service-concept derivation edges w.r.t. $s_t$, $(s_t \delta^{s \to c} c_{s_t}) \in E_O$ **do**
16:         # prune if elements in $\Phi$ do not exist in $c_{s_t}$'s derivation path, that is, the
           union of all its parents' $\psi$ graphs
17:         **if** $(\Phi \subseteq \bigcup \psi(c_{s_t}))$ **then**
18:             $W' \leftarrow$ WFEnum_key($c_{s_t}$, $\Phi \cap \psi(c_{s_t})$, $Q$, $W$, $O$)
19:             **if** $W' \neq ()$ **then**
20:                 $W_{s_t} \leftarrow W_{s_t} \times W'$
21:                 $W \leftarrow W \cup W'$
22:             **end if**
23:         **end if**
24:     **end for**
25:     # construct service invocation plan for each $p \in W_{s_t}$, and append to $W$
26:     **for all** $p \in W_{s_t}$ **do**
27:         $W \leftarrow W \cup \{(s_t, p)\}$
28:     **end for**
29: **end for**
30: return $W$

---

*img2*, also derives the *image* concept. Assuming that $\Phi = \{shore\}$, because the $\psi$-graphs pertaining to all of *img2*'s parameters does not account for the elements in $\Phi$, *img2* can be immediately pruned here (Line 17). Otherwise, service $s_t$ is deemed promising, and its parameters' concepts are used as targets to generate workflow (sub)plans toward the total realization of $s_t$. Recalling the workflow's recursive definition from previously, this step is tantamount to deriving the nonterminal case where $(s_t, (w_1, \ldots, w_p)) \in S$. Finally, whereas the complete plan for $s_t$ is included in the result set (Line 27), $W$, each (sub)plan is also included because they include some subset of $\Phi$, the required keyword concepts, and therefore could be somewhat relevant to the user's query (Line 21).

---

**Algorithm 3** KMQuery($K$, $O$)

---

1: $R \leftarrow ()$ # $R$ will hold the list of derived workflow results
2: $Q_K \leftarrow O.\text{mapParams}(K)$
3: $C_K \leftarrow O.\text{mapConcepts}(K \setminus Q_K)$
4: # compute the power set, $\mathcal{P}(C_K)$, of $C_K$
5: **for all** $\rho \in \mathcal{P}(C_K)$, in descending order of $|\rho|$ **do**
6:     # $\rho = \{c_1, \ldots, c_n\}, \{c_1, \ldots, c_{n-1}\}, \ldots, \{c_1\}$
7:     # check for reachability within $\rho$, and find successor if true
8:     *reachable* $\leftarrow$ false
9:     **for all** $c_i \in \rho \wedge \neg reachable$ **do**
10:        **if** $(\rho \setminus \{c_i\}) \subseteq \psi(c_i)$ **then**
11:            $c_{\text{root}} \leftarrow c_i$
12:            *reachable* $\leftarrow$ true
13:        **end if**
14:     **end for**
15:     **if** *reachable* **then**
16:        # from ontology, enumerate all plans with $c_{\text{root}}$ as target
17:        $R \leftarrow R \cup \text{WFEnum\_key}(c_{\text{root}}, (\rho \setminus \{c_{\text{root}}\}), Q_K, O)$
18:        # prune all subsumed elements from $\mathcal{P}(C_K)$
19:        **for all** $\rho' \in \mathcal{P}(C_K)$ **do**
20:            **if** $\rho' \subseteq \rho$ **then**
21:                $\mathcal{P}(C_K) \leftarrow \mathcal{P}(C_K) \setminus \{\rho'\}$
22:            **end if**
23:        **end for**
24:     **end if**
25: **end for**
26: return $R$

---

With the planning algorithm in place, the natural extension now is to determine its input from a given list of keywords.

The query planning algorithm, shown in Algorithm 3, simply takes a set of keywords, $K$, and the ontology, $O$, as input, and the resulting list of workflow plans, $R$, is returned. First, the set of query parameters, $Q_K$, is identified using the concept pattern mapper on each of the key terms. Because user-issued parameter values are essentially data, they define a $\delta^{c \rightarrow d}$-type derivation on the concepts to which they are mapped. Here, $(longitude \delta^{c \rightarrow d} x)$, $(latitude \delta^{c \rightarrow d} y)$, $(date \delta^{c \rightarrow d} 8/20/2009)$, can be identified as a result (Line 2). The remaining concepts from $K$ are also determined, $C_K = \{wind, shore, image, coastal\text{-}terrain\text{-}model\}$ (note that "coast" had been deduced to the concept *shore* and that "line" had been dropped since it did not match any concepts in $O$).

Next (Lines 5–14), the algorithm attempts to plan workflows incorporating all possible combinations of concepts within $C_K$. The power set, $\mathcal{P}(C_K)$ is computed for $C_K$, to contain the set of all subsets of $C_K$. Then, for each subset-element $\rho \in \mathcal{P}(C_K)$, the algorithm attempts to find the root concept in the derivation graph

produced by $\rho$. For example, when $\rho = \{shore, image, coastal\text{-}terrain\text{-}model\}$, the
root concept is *image* in Fig. 5.4. However, when $\rho = \{shore, coastal\text{-}terrain\text{-}model\}$, then $c_{root} = shore$. But since any workflows produced by the former sub-
sumes any produced by the latter $\rho$ set of concepts, the latter can be pruned (thus
why we loop from descending order of $|\rho|$ on Line 5). In order to perform the root-
concept test, for each concept element $c_i \in \rho$, its $\psi$-graph $\psi(c_i)$ is first computed,
and if it consumes all other concepts in $\rho$, then $c_i$ is determined to be the root (recall
that $\psi(c_i)$ generates a concept-derivation DAG rooted in $c_i$).

Back to our example, although *wind* is a valid concept in $O$, it does not con-
tribute to the derivation of any of the relevant elements. Therefore, when $\rho = \{wind, image, shore, coastal\text{-}terrain\text{-}model\}$, no plans will be produced because *wind* is
never reachable regardless of which concepts is considered root. The next $\rho$, how-
ever, produces $\{image, shore, coastal\text{-}terrain\text{-}model\}$. Here, $\psi(image)$ incorporates
both *shore* and *coastal-terrain-model*, and thus, *image* is determined to be $c_{root}$. The
inner loop on Line 9 can stop here, because the DAG properties of $O$ does not permit
$\psi(shore)$ or $\psi(coastal\text{-}terrain\text{-}model)$ to include *shore*, and therefore neither can be
root for this particular $\rho$.

When a reachable $\rho$ subset has been determined, the planning method,
WFEnum_key can be invoked (Lines 15–24). Using $c_{root}$ as the targeted with
$\rho \setminus \{c_{root}\}$ being the concepts required in the derivation paths toward $c_{root}$,
WFEnum_key is employed to return all workflow plans. But as we saw in Al-
gorithm 1, WFEnum_key also returns any workflow (sub)plans that were used to
derive the target. That is, although *image* is the target here, the *shore* concept would
have to be first derived to substantiate it, and it would thus be included in $R$ as a
separate plan. Due to this redundancy, after WFEnum_key has been invoked, Lines
18–23 prunes the redundant $\rho$'s from the power set. In our example, every subset
element will be pruned except when $\rho = \{wind\}$. Therefore, *wind* would become
rooted its workflows will likewise be planned separately.

### 5.4.3 Relevance Ranking

The resulting workflow plans should be ordered by their relevance. Relevance, however, is a somewhat loose term under our context. We define relevance as a function
of the number of keyword-concepts that appear in each workflow plan. We, for instance, would expect that any workflow rooted in *wind* be less relevant to the user
than the plans which include significantly more keyword-concepts: *shore*, *image*,
etc. Given a workflow plan $w$ and query $K$, we measure $w$'s relevance score, as
follows:

$$r(w, K) = \frac{|V_\psi(w) \cap C(K)|}{|C(K)| + \log(|C(K) \setminus V_\psi(w)| + 1)}.$$

Recall that $V_\psi(w)$ denotes the set of concept vertices in $w$'s concept derivation
graph, $\psi(w)$. Here, $C(K)$ represents the set of concept nodes mapped from $K$.

This equation corresponds to the ratio of the amount of concepts from $C(K)$ that $w$ captures. The log term in the denominator signifies a slight *fuzziness* penalty for each concept in $w$'s derivation graph that was not specified in $K$. The motivation for this penalty is to reward "tighter" workflow plans are that more neatly represented (and thus, more easily understandable and interpreted by the user). This metric is inspired by traditional approaches for answering keyword queries over relational databases [1, 43].

## 5.4.4 A Case Study

We present a case study of our keyword search functionality in this section. Our system is run on an Ubuntu Linux machine with a Pentium 4 3.00 GHz Dual Core with 1 GB of RAM. This work has been a cooperative effort with the Department of Civil and Environmental Engineering and Geodetic Sciences here at the Ohio State University. Our collaborators supplied us with various services that they had developed to process certain types of geospatial data. A set of geospatial data was also given to us. In all, the ontology used in this experiment consists of 29 concepts, 25 services, 5 data types. The 25 services and 2248 data files were registered to the ontology based on their accompanying metadata, solely for the purposes of this experiment. We note that, although the resource size is small, the given is sufficient for evaluating the *functionality* of keyword search support. A set of queries, shown in Table 5.1, are used to evaluate our system.

First, we present the search time of the six queries issued to the system. In this experiment, we executed the search using two versions of our algorithm. Here, the search time is the sum of the runtimes for KMQuery and WFEnum_key algorithms. The first version consists of the a priori pruning logic, and the second version does not prune until the very end. The results of this experiment are shown in Fig. 5.5, and as we can see, a typical search executes on the order of several milliseconds, albeit that the ontology size is quite small.

We can also see that the pruning version results in slightly faster search times in almost all queries, with the exception of QueryID $= 3$. It was later verified that this query does not benefit from pruning with the given services and data sets. In other

**Table 5.1** Experimental queries

| Query ID | Description |
| --- | --- |
| 1 | "coast line CTM 7/8/2003 (41.48335,–82.687778)" |
| 2 | "bluff line DEM 7/8/2003 (41.48335,–82.687778)" |
| 3 | "(41.48335,–82.687778) 7/8/2003 wind CTM" |
| 4 | "waterlevel=174.7 cm water surface 7/8/2003 (41.48335,–82.687778)" |
| 5 | "waterlevel (41.48335,–82.687778) 13:00:00 3/3/2009" |
| 6 | "land surface change (41.48335,–82.687778) 7/8/2003 7/7/2004" |

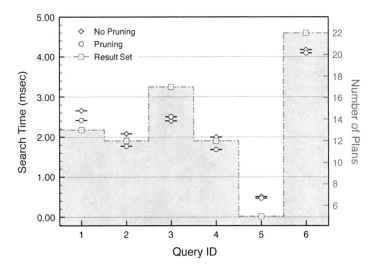

**Fig. 5.5** Search time

words, the pruning logic is an overhead for this case. Along the right $y$-axis, the result set size is shown. Because the test data set is given by our collaborators, in addition to the fact that our search algorithm is exhaustive, we can claim (and it was later verified) that the recall is 100%. Recall by itself, however, is not sufficient to measuring the effectiveness of the search.

To measure the precision of the result set, we again required the help of our collaborators. For each workflow plan $w$ in the result set, the domain experts assigned a score $r'(w, K)$ from 0 to 1. The precision for each plan is then measured relative to the difference of this score to the relevance score $r(w, K)$ assigned by our search engine. For a result set $R$, its precision is thus computed,

$$\text{prec}(R, K) = \frac{1}{|R|} \sum_{w \in R} 1 - \left( \left| r(w, K) - r'(w, K) \right| \right)$$

The precision for our queries is plotted in Fig. 5.6. Most of the variance are introduced due to the fact that our system underestimated the relevance of some plans. Because Query 3 appeared to have performed the worst, we show its results in Table 5.2. The third query contains five concepts after keyword-concept mapping: *wind*, *date*, *longitude*, *latitude*, and *coastal-terrain-model*. The first five plans enumerated captures all five concepts plus "water surface," which is superfluous to the keyword query. Therefore, any plans generating a water surface will be slightly penalized. Note that, while the variance is relatively high when compared with the user's expectations, the scores do not affect the user's expected overall ordering of the results. Although, it certainly can be posited that other properties, such as cost/quality of the workflow, can be factored into the relevance calculation.

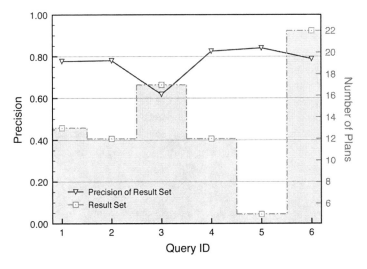

**Fig. 5.6**  Precision of search results

**Table 5.2**  QueryID 3 results set and precision

| Workflow plan | $r$ | $r'$ |
|---|---|---|
| GetWindVal(GetWaterSurface(getCTMLowRes(CTM42.dat))) | 0.943 | 0.8 |
| GetWindVal(GetWaterSurface(getCTMMedRes(CTM42.dat))) | 0.943 | 0.8 |
| GetWindVal(GetWaterSurface(getCTMHighRes(CTM42.dat))) | 0.943 | 0.8 |
| GetWindVal(GetWaterSurface(CreateFromUrlLowRes(CTM42.dat))) | 0.943 | 0.8 |
| GetWindVal(GetWaterSurface(CreateFromUrlHighRes(CTM42.dat))) | 0.943 | 0.8 |
| getCTMLowRes(CTM42.dat) | 0.8 | 0.3 |
| getCTMMedRes(CTM42.dat) | 0.8 | 0.3 |
| getCTMHighRes(CTM42.dat) | 0.8 | 0.3 |
| CreateFromUrlLowRes(CTM42.dat) | 0.8 | 0.3 |
| CreateFromUrlHighRes(CTM42.dat) | 0.8 | 0.3 |
| CTM42.dat | 0.8 | 0.3 |
| GetWaterSurface(getCTMLowRes(CTM42.dat)) | 0.755 | 0.3 |
| GetWaterSurface(getCTMMedRes(CTM42.dat)) | 0.755 | 0.3 |
| GetWaterSurface(getCTMHighRes(CTM42.dat)) | 0.755 | 0.3 |
| GetWaterSurface(CreateFromUrlLowRes(CTM42.dat)) | 0.755 | 0.3 |
| GetWaterSurface(CreateFromUrlHighRes(CTM42.dat)) | 0.755 | 0.3 |

## 5.5 Experimental Results

In this section, we discuss the performance evaluation for two aspects we described
in Sect. 5.3, namely, (i) QoS handling and (ii) the benefits afforded by our caching
framework in an actual cloud environment, particularly in conjunction with elas-

ticity available in cloud environments. For both experiments, we use the coast line extraction query seen throughout this paper. We have evaluated these two features as we believe they are unique to the Auspice system. Referring back to Fig. 5.3, the coast line workflow is composed of two services: (1) extractShoreline, which inputs a water level reading and a coastal terrain model (CTM) data file; (2) The parent service to extractShoreline is getWaterLevel, which inputs the time and location of interest from the user query. Whereas the water level service is negligible due to parallelism, the actual coast line extraction service is time consuming, i.e., CTM files are quite large.

### 5.5.1 QoS Handling

In this experiment, we allow the user to specify the amount accuracy they require and report Auspice's efforts on meeting them. The execution time of this workflow can be shortened by sampling the CTM input file. However, as we illustrated in Sect. 5.3.1, the sampling rate of CTMs can take quite a departure from the domain-specific accuracies that the users specify. A model may predict actual errors (in meters) of the extractShoreline service operation on varying CTM sampling rates, but as for planning, Auspice must inversely determine the sampling rate on the input CTM from this model. Table 5.3 shows Auspice's efforts toward making these decisions on a particular CTM file.

The left half of the table shows the correct correspondences between the sampling rate of the CTM ($\alpha\%$) and the physical errors predicted to have been induced by the loss in resolution. The right half of the table shows Auspice's suggested $\alpha$ and the corresponding errors. Seen in the tables' juxtaposition, Auspice's automatic suggestions for $\alpha$ come very close to the ideal values and contribute insignificantly to the differences in physical errors. Although not shown here, the time taken for

**Table 5.3** Auspice suggested sampling rates

| Ideal | | Suggested | |
|---|---|---|---|
| $\alpha\%$ | Error (meters) | $\alpha\%$ | Error (meters) |
| 10 | 61.1441 | 10.00 | 61.1441 |
| 20 | 30.7205 | 19.93 | 30.7204 |
| 30 | 20.4803 | 29.91 | 20.4798 |
| 40 | 15.3603 | 39.89 | 15.3599 |
| 50 | 12.2882 | 49.87 | 12.2892 |
| 60 | 10.2402 | 59.98 | 10.2392 |
| 70 | 8.7773 | 69.88 | 8.7769 |
| 80 | 7.6801 | 79.90 | 7.6803 |
| 90 | 6.8268 | 89.94 | 6.8266 |
| 100 | 6.1441 | 100 | 6.1441 |

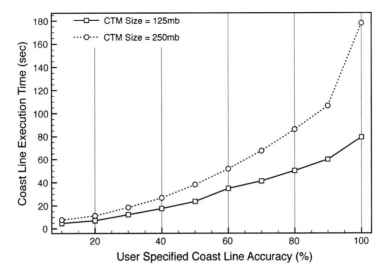

**Fig. 5.7**  QoS handling in coast line extraction

suggesting $\alpha$ is on the order of $10^{-6}$ seconds. Figure 5.7 displays the total execution times of the coast line extraction workflow on the user given accuracies (along the $x$-axis).

## 5.5.2 Caching in a Cloud Environment

We employ the Amazon Elastic Compute Cloud (EC2) [3] to support all of our experiments. Each Cloud node instance runs an Ubuntu Linux image on which our cache server logic is installed. Each image runs on a *Small EC2 Instance*, which, according to Amazon, comprises 1.7 GB of memory, 1 virtual core (equivalent to a 1.0–1.2 GHz 2007 Opteron or 2007 Xeon processor) on a 32-bit platform. In all of our experiments, the caches are initially cold.

We executed repeated runs of the shoreline extraction query. The baseline execution time of the composite services, i.e., when executed without any caching, takes approximately ~23 seconds to complete. Because waiting for each request to complete over the lifetime of the experiment would take prohibitive amounts of time, we simulated its execution. The inputs to each shoreline extraction query consist the desired location and date. We have randomized these inputs over 64K possibilities for each request. The randomized query requests emulates the worst-case scenario.

The initial experiment evaluates the effects of the cache without node contraction. In other words, the length of our eviction sliding window is ∞. Under this configuration, our cache is able to grow as large as it needs to handle the size of the cache. We run our cache system over static, fixed-node configurations (*static-2, static-4, static-8*). We then compare these static versions against our dynamic algorithm, Greedy Bucket Allocation (*GBA*), which runs over the EC2 public Cloud.

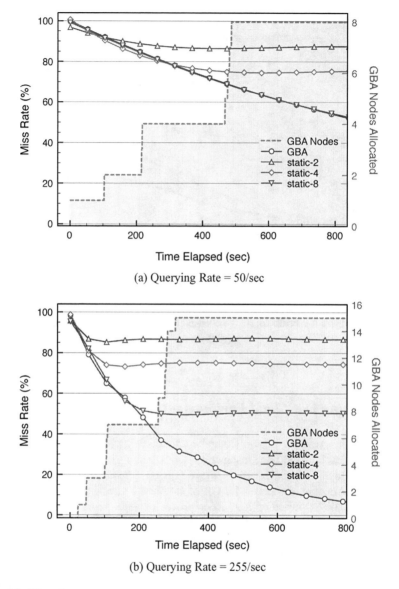

(a) Querying Rate = 50/sec

(b) Querying Rate = 255/sec

**Fig. 5.8** Miss rate

We executed the shoreline mashup at a rate of 50 query requests per second and 255 query requests per second. Figures 5.8a and b display the miss rates over the span of 800 seconds under these two configurations. Notice that the miss rates (against the left $y$-axis) for *static-2*, *static-4*, and *static-8* converge at relatively high values somewhat early into the experiment due to capacity being reached, although this behavior is exposed for *static-8* only in Fig. 5.8b. Because we are executing *GBA* with an infinite eviction window, we do not encounter this capacity issue. In fact, our method continues to improve miss rates beyond the static versions, albeit requiring more nodes to handle the queries. Toward the end of the run, *GBA* is capable of attaining near-zero miss rates.

The node allocation behavior (against the right $y$-axis) shows that, over the lifespan of this experiment, *GBA* only employs a fraction of the nodes needed to perform better than *static-8*. In fact, an average of $\lceil 5.557 \rceil = 6$ nodes is used over the query rate $= 50$ experiment and an average of $\lceil 12.6 \rceil = 13$ nodes in the query rate $= 255$ version. This translates to less overall EC2 usage cost per performance over static allocations. The growth of nodes is also not unexpected, though, at first glance it appears to be exponential. Node increases are concentrated toward the beginning of execution because the overall capacity is too small to handle the query rate.

Figures 5.9a and b, which show the respective mean query execution times, correspond directly to the above miss rates. To create these figures, for each second elapsed in our execution, we averaged the query execution times (over 50 and 255 respectively) and plotted them (against the left $y$-axis). The speedup provided by the static versions expectedly flatten somewhat quickly, again, due to the nodes reaching capacity. *GBA*, on the other hand, performed better, but requiring far less nodes throughout the length of the experiment.

To show node allocation and migration overhead, in Figs. 5.10a and b, the maximum query execution time (*GBA-max*) is displayed for *GBA*. The spikes in the execution time expectedly align with EC2 instance allocation which, in our experience, can take extensive amounts of time. We posit that the demand for node allocation diminishes as the experiment proceeds even with high querying rates. But while node allocation overheads are high, its negative impact on overall speedup is amortized over the span of the experiment because it is only seldom invoked. Moreover, techniques, such as preloading EC2 instances, can also be used to further minimize this overhead although these have not been considered in this paper.

Next, we evaluated the sliding window approach of maintaining our cache. Two separate experiments were devised to show the effectiveness of the sliding window and to show that our cache is capable of scaling down. Again, we randomize the query inputs over 64K possibilities. We begin these experiments with a querying rate of 25/s. Then between 100 and 300 seconds, the querying rate is increased to 60/s, and input possibilities are decreased down to 32K. This emulates a period of frequent and highly related queries being issued. After 300 seconds into the experiment, the querying rate resumes back to 25/s with 64K possible inputs.

Figures 5.11a and b show the results of this experiment for sliding window sizes of 50 sec and 100 sec respectively. Specifically, the cache will attempt to maintain, with high probability, all records that were queried in the most recent 50 and 100 seconds. The decay $\alpha$ has been fixed at 0.99 for these experiments. The eviction threshold $T_\lambda$ is set at the baseline $\alpha^{m-1} \approx 0.367$ to evict any key which had been only queried within the evicted slice.

As demonstrated by these experiments, our cache adapts to the query intensive period by lowering mean query execution times. We can also observe that, after the query intensive period expires at 300 seconds, the sliding window detects the normal querying rates and removes nodes as they become superfluous (though this is only

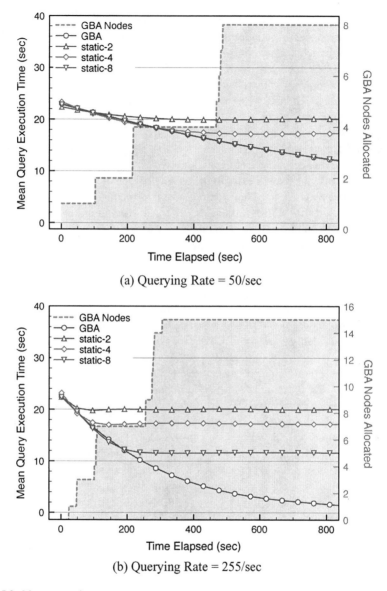

(a) Querying Rate = 50/sec

(b) Querying Rate = 255/sec

**Fig. 5.9** Mean query times

noticeable in Fig. 5.11b. Here, the nodes do not decrease back down to one because our contraction algorithm is slightly conservative: Recall that we only contract if two of the lowest loaded nodes account for less than half the space required to store their merged cache.

The differences between the two side-by-side figures also imply that the size of the sliding window is a determinant factor on both performance and cost. Since the

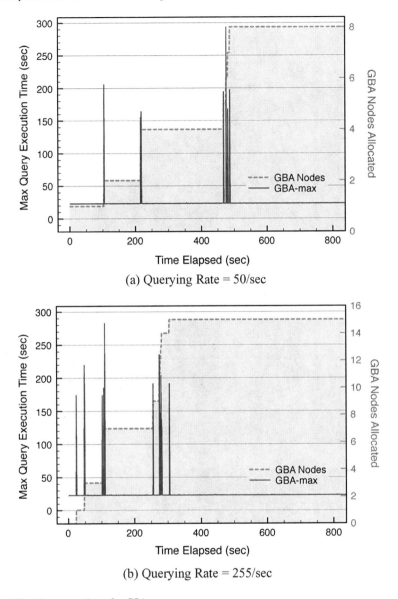

(a) Querying Rate = 50/sec

(b) Querying Rate = 255/sec

**Fig. 5.10** Max query times for *GBA*

querying rate and sliding window are so small, this version never requires more than one node, and it requires less nodes than its counterpart in Fig. 5.11b. However, the tradeoff is clear in that it cannot reap the benefits of the larger cache afforded by the longer sliding window. The same experiments were conducted for higher querying rates to produce Figs. 5.12a and b. We increased the normal querying rate to 50/s and intensive querying rate to 255/s. As seen in the figures, the same cache elasticity

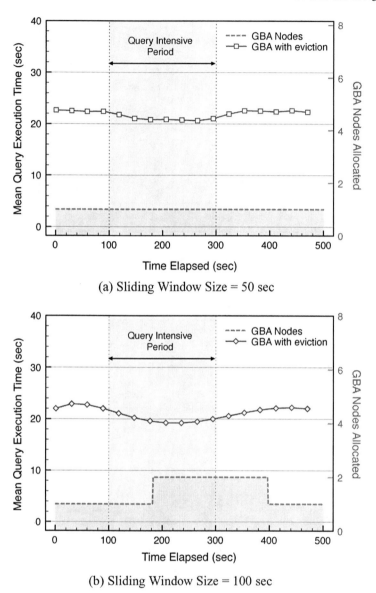

(a) Sliding Window Size = 50 sec

(b) Sliding Window Size = 100 sec

**Fig. 5.11** Cache contraction (Normal Query Rate = 25/s, Intensive Rate = 60/s)

behavior can be expected upon these very high rates of querying, showing us some positive results on scalability. In fact, in terms of performance our system welcomes higher querying rates, as it populates our cache more frequently within the sliding window.

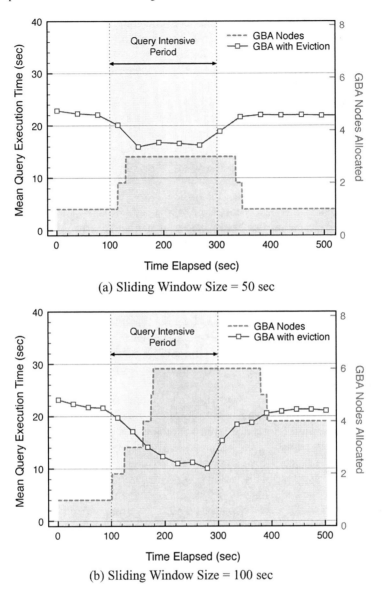

(a) Sliding Window Size = 50 sec

(b) Sliding Window Size = 100 sec

**Fig. 5.12** Cache contraction (Normal Query Rate = 50/s, Intensive Rate = 255/s)

## 5.6 Related Works

Among the first systems to utilize workflows to manage scientific processes is ZOO [32], which employs an object-oriented language to model the invocation of processes and the relationships between them. Another notable system, Condor [37], was originally proposed to harvest the potential of idle CPU cycles. Soon after, dependent processes (in the form of directed acyclic graphs) were being scheduled on Condor systems using DAGMan [17]. Recently, with the onset of the Data Grid, Condor has been integrated with the Globus Tookit [24] into Condor-G [25]. Pegasus [19] creates workflow plans in the form of Condor DAGMan files, which then uses the DAGMan scheduler for execution.

Many systems currently allow users to guide workflow designs. In Casati et al.'s eFlow [10], a workflow's structure is defined by users, but the instantiation of services within the process is dynamically allocated by the eFlow engine. Sirin et al. proposed a user interactive composer that provides semi-automatic composition [42]. In their system, after each time that a particular service is selected for use in the composition, the user is presented a filtered list of possible choices for the next step. This filtering process is made possible by associating semantic data with each service. To complement the growing need for interoperability and accessibility, many prominent workflow managers, including Pegasus [19], Taverna [40], Kepler [2] (the service-enabled successor to the actor and component-based Ptolemy II [9]), and Triana [38] have become attuned with service-oriented systems.

Workflow systems with QoS support have also been developed. For example, Askalon offers system-level QoS, e.g., throughput and transfer rates [7, 8]. Glatard's service scheduler exploits parallelism within service and data components [31]. Lera et al. have developed a performance ontology for dynamic QoS assessment [36]. Kumar et al. [35] have developed a framework for application-level QoS support. Their system, which integrates well-known Grid utilities (the Condor scheduler [25], Pegasus [19], and DataCutter [6], a tool which enables pipelined execution of data streams) considers *quality-preserving* (e.g., chunk size manipulation, which does not adversely affect accuracy of workflow derivations) and *quality-trading* QoS parameters (e.g., resolution, which could affect one QoS in order to optimize another). In quality-preserving support, the framework allows for parameters, such as chunksize, to be issued. These types of parameters have the ability to modify a workflow's parallelism and granularity, which potentially reduces execution times without performance tradeoffs. For quality-trading QoS support, an extension to the Condor scheduler implements the tradeoff between derived data accuracy for improved execution time. We believe that Auspice is the first to propose algorithms reconciling the tradeoffs between data reduction parameters against actual domain errors within the scientific application.

## 5.7 Conclusion

At the Ohio State University, we have developed Auspice, an automatic service composition engine that supports several functionalities, outlined below.

- Enables a nonintrusive framework for sharing scientific data sets and Web services through metadata indexing.
- Composes known services and data sets in disparate ways to derive user queries.
- Through error models, it adaptively controls tradeoffs between execution time and application error to meet QoS constraints.

- Caches intermediate results for trivializing future service invocations.

We are currently exploring several directions to further our system's functionalities. We will finalize the keyword search process and also, inspect methods toward the seamless integration of Deep Web data sources. We also believe that the emergence of the Cloud affords us excellent opportunities to develop fruitful research. On this front we propose to examine the effects of parallelism in a Cloud versus static systems, such as a cluster. It is well known that one way to reduce execution time is by parallelizing large, independent computations within scientific workflow systems [35]. While it is tempting to maximize the usage of the Cloud's "infinite resources," overheads, such as deploying virtual machine images onto new nodes, exist. For disparate applications and data sets, there will certainly be cases where the Cloud's overheads are amortized against the execution time speed ups. Conversely, there will also be applications where the Cloud's overheads override the benefits for its use. We believe that a study on workloads, parallelism granularity, and cost would be beneficial toward understanding the tradeoffs between Cloud usage and other preexisting infrastructures.

**Acknowledgements**  This work is supported by NSF grants 0541058, 0619041, and 0833101. The equipment used for the experiments reported here was purchased under the grant 0403342.

# References

1. Agrawal, S.: Dbxplorer: a system for keyword-based search over relational databases. In: ICDE, pp. 5–16 (2002)
2. Altintas, I., Berkley, C., Jaeger, E., Jones, M., Ludscher, B., Mock, S.: Kepler: an extensible system for design and execution of scientific workflows (2004)
3. Amazon elastic compute cloud. http://aws.amazon.com/ec2
4. Armbrust, M., et al.: Above the clouds: a Berkeley view of cloud computing. Technical Report UCB/EECS-2009-28, EECS Department, University of California, Berkeley, Feb 2009
5. The atlas experiment. http://atlasexperiment.org
6. Beynon, M.D., Kurc, T., Catalyurek, U., Chang, C., Sussman, A., Saltz, J.: Distributed processing of very large datasets with datacutter. Parallel Comput. **27**(11), 1457–1478 (2001)
7. Brandic, I., Pllana, S., Benkner, S.: An approach for the high-level specification of QoS-aware grid workflows considering location affinity. Sci. Program. **14**(3–4), 231–250 (2006)
8. Brandic, I., Pllana, S., Benkner, S.: Specification, planning, and execution of QoS-aware grid workflows within the Amadeus environment. Concurr. Comput. Pract. Exp. **20**(4), 331–345 (2008)
9. Brooks, C., Lee, E.A., Liu, X., Neuendorffer, S., Zhao, Y., Zheng, H.: Heterogeneous concurrent modeling and design in Java (vol. 2: Ptolemy II software architecture). Technical Report 22, EECS Dept., UC Berkeley, July 2005
10. Casati, F., Ilnicki, S., Jin, L., Krishnamoorthy, V., Shan, M.-C.: Adaptive and dynamic service composition in eFlow. In: Conference on Advanced Information Systems Engineering, pp. 13–31 (2000)
11. Chiu, D., Agrawal, G.: Enabling ad hoc queries over low-level scientific datasets. In: Proceedings of the 21th International Conference on Scientific and Statistical Database Management (SSDBM'09) (2009)
12. Chiu, D., Agrawal, G.: Flexible caches for derived scientific data over cloud environments. Technical Report OSU-CISRC-7/09-TR35, Department of Computer Science and Engineering, The Ohio State University, July 2009

13. Chiu, D., Agrawal, G.: Hierarchical caches for grid workflows. In: Proceedings of the 9th IEEE International Symposium on Cluster Computing and the Grid (CCGRID). IEEE, New York (2009)
14. Chiu, D., Deshpande, S., Agrawal, G., Li, R.: Composing geoinformatics workflows with user preferences. In: Proceedings of the 16th ACM SIGSPATIAL International Conference on Advances in Geographic Information Systems (GIS'08), New York, NY, USA (2008)
15. Chiu, D., Deshpande, S., Agrawal, G., Li, R.: Cost and accuracy sensitive dynamic workflow composition over grid environments. In: Proceedings of the 9th IEEE/ACM International Conference on Grid Computing (Grid'08) (2008)
16. Chiu, D., Deshpande, S., Agrawal, G., Li, R.: A dynamic approach toward QoS-aware service workflow composition. In: Proceedings of the 7th IEEE International Conference on Web Services (ICWS'09). IEEE Computer Society, Los Alamitos (2009)
17. Condor dagman. http://www.cs.wisc.edu/condor/dagman
18. Dublin core metadata element set, version 1.1 (2008)
19. Deelman, E., Singh, G., Su, M.-H., Blythe, J., Gil, Y., Kesselman, C., Mehta, G., Vahi, K., Berriman, G.B., Good, J., Laity, A.C., Jacob, J.C., Katz, D.S.: Pegasus: a framework for mapping complex scientific workflows onto distributed systems. Sci. Program. **13**(3), 219–237 (2005)
20. Fellbaum, C. (ed.): WordNet: An Electronic Lexical Database. MIT Press, Cambridge (1998)
21. Metadata ad hoc working group.content standard for digital geospatial metadata (1998)
22. Federal geospatial data clearinghouse. http://clearinghouse.fgdc.gov
23. Foster, I.: Service-oriented science. Science **308**(5723), 814–817 (2005)
24. Foster, I., Kesselman, C.: Globus: a metacomputing infrastructure toolkit. Int. J. Supercomput. Appl. **11**, 115–128 (1996)
25. Frey, J., Tannenbaum, T., Foster, I., Livny, M., Tuecke, S.: Condor-G: a computation management agent for multi-institutional grids. In: Proceedings of the Tenth IEEE Symposium on High Performance Distributed Computing (HPDC), San Francisco, CA, August 2001, pp. 7–9 (2001)
26. gbio: Grid for bioinformatics. http://gbio-pbil.ibcp.fr
27. Bioinfogrid. http://www.bioinfogrid.eu
28. Biomedical informatics research network. http://www.nbirn.net
29. Cyberstructure for the geosciences. http://www.geongrid.org
30. The geography network. http://www.geographynetwork.com
31. Glatard, T., Montagnat, J., Pennec, X.: Efficient services composition for grid-enabled data-intensive applications (2006)
32. Ioannidis, Y.E., Livny, M., Gupta, S., Ponnekanti, N.: Zoo: a desktop experiment management environment. In: VLDB '96: Proceedings of the 22th International Conference on Very Large Data Bases, pp. 274–285. Morgan Kaufmann, San Francisco (1996)
33. Jensen, C.S., Lin, D., Ooi, B.C.: Query and update efficient $B^+$-tree based indexing of moving objects. In: Proceedings of Very Large Databases (VLDB), pp. 768–779 (2004)
34. Karger, D., et al.: Consistent hashing and random trees: distributed caching protocols for relieving hot spots on the world wide web. In: ACM Symposium on Theory of Computing, pp. 654–663 (1997)
35. Kumar, V.S., Sadayappan, P., Mehta, G., Vahi, K., Deelman, E., Ratnakar, V., Kim, J., Gil, Y., Hall, M., Kurc, T., Saltz, J.: An integrated framework for performance-based optimization of scientific workflows. In: HPDC '09: Proceedings of the 18th ACM International Symposium on High Performance Distributed Computing, pp. 177–186. ACM, New York (2009)
36. Lera, I., Juiz, C., Puigjaner, R.: Performance-related ontologies and semantic web applications for on-line performance assessment intelligent systems. Sci. Comput. Program. **61**(1), 27–37 (2006)
37. Litzkow, M., Livny, M., Mutka, M.: Condor—a hunter of idle workstations. In: Proceedings of the 8th International Conference of Distributed Computing Systems, June 1988
38. Majithia, S., Shields, M.S., Taylor, I.J., Wang, I.: Triana: a graphical web service composition and execution toolkit. In: Proceedings of the IEEE International Conference on Web Services (ICWS'04), pp. 514–524. IEEE Computer Society, Los Alamitos (2004)

39. Biological data working group. Biological data profile (1999)
40. Oinn, T., Addis, M., Ferris, J., Marvin, D., Senger, M., Greenwood, M., Carver, T., Glover, K., Pocock, M.R., Wipat, A., Li, P.: Taverna: a tool for the composition and enactment of bioinformatics workflows. Bioinformatics **20**(17), 3045–3054 (2004)
41. Sloan digital sky survey. http://www.sdss.org
42. Sirin, E., Parsia, B., Hendler, J.: Filtering and selecting semantic web services with interactive composition techniques. IEEE Intell. Syst. **19**(4), 42–49 (2004)
43. University, V.H., Hristidis, V.: Discover: keyword search in relational databases. In: VLDB, pp. 670–681 (2002)
44. Wan, M., Rajasekar, A., Moore, R., Andrews, P.: A simple mass storage system for the SRB data grid. In: MSS '03: Proceedings of the 20th IEEE/11th NASA Goddard Conference on Mass Storage Systems and Technologies (MSS'03), p. 20. IEEE Computer Society, Washington (2003)
45. Extensible markup language (xml) 1.1 (second edition)

# Chapter 6
# Parameter Sweep Job Submission to Clouds

**P. Kacsuk, A. Marosi, M. Kozlovszky, S. Ács,
and Z. Farkas**

**Abstract** This chapter introduces the existing connectivity and interoperability issues of Clouds, Grids, and Clusters and provides solutions to overcome these issues.
The paper explains the principles of parameter sweep job execution by P-GRADE
portal and gives some details on the concept of parameter sweep job submission
to various Grids by the 3G Bridge. Then it proposes several possible solution variants how to extend the parameter sweep job submission mechanism of P-GRADE
and 3G Bridge toward Cloud systems. Finally, it shows the results of performance
measurements that were gained for the proposed solution variants.

## 6.1 Introduction

The e-Science infrastructure ecosystem has been recently enriched with Clouds,
and, as the result, the main pillars of this ecosystem are:

- Supercomputer-based grids (e.g., DEISA, TeraGrid, etc.)
- Cluster-based Grids (e.g., EGEE, NorduGrid, OSG, SEE-GRID, etc.)
- Desktop Grids (e.g., BOINC-based, XtremWeb-based, etc.)
- Clouds (e.g., Eucalyptus, OpenNebula, Amazon, etc.)

All pillars have their own advantages that make them attractive for a certain application area. Supercomputers are very reliable, robust services that can be used
to solve tightly coupled compute- or data-intensive applications. Their drawback is
the high investment and maintenance cost that is affordable only for a small number of distinguished computing centers that typically receive financial support from
national or regional governments. Organizing them into a Grid system where large
number of scientists can access them in a balanced and managed way significantly
increases their utilization.

P. Kacsuk (✉) · A. Marosi · M. Kozlovszky · S. Ács · Z. Farkas
MTA SZTAKI, P.O. Box 63, 1518 Budapest, Hungary
e-mail: kacsuk@sztaki.hu

M. Cafaro, G. Aloisio (eds.), *Grids, Clouds and Virtualization,*
Computer Communications and Networks,
DOI 10.1007/978-0-85729-049-6_6, © Springer-Verlag London Limited 2011

Cluster-based service Grids are less expensive than Supercomputer-based grids. They are very flexible in the sense that they can efficiently run any kind of applications including tightly and loosely coupled, compute- and data-intensive applications. These are the most popular forms of building Grid systems. Nowadays Grid computing is used in many research domains to provide the required large set of computing and storage resources for e-scientists. Europe's leading Grid computing project Enabling Grids for E-sciencE (EGEE) is a good example, providing a computing support infrastructure for over 13,000 researchers world-wide, from fields as diverse as high-energy physics, chemistry, engineering, earth and life sciences.

Desktop Grids represent the least expensive form of collecting resources. The two main options are the global volunteer computing systems and the institutional Desktop Grid (DG) systems. In the global version the spare cycles of typically home computers are donated on a volunteer basis. In DG systems, anyone can bring resources into the grid system, offering them for the common goal of that Grid. Installation and maintenance of the grid resource middleware is extremely simple, requiring no special expertise. This ease of use enables large numbers of donors to contribute into the pool of shared resources. In the institutional DG systems the spare cycles of the existing computers of an (academic or commercial) institution are exploited. Since it uses the free cycles of already available computers, it needs only minimal initial investment and maintenance cost. Volunteer DG systems are very popular and collect very large number of resources in the range of 10K–1M CPUs. The most well-known example of an application that has successfully adapted to DGs is the SETI@HOME project, in which approximately four million PCs have been involved. However, the drawback of Desktop Grids is that they are not suitable for all kind of applications. They can efficiently support only bag of task (parameter sweep and master/worker) compute-intensive applications, however they cannot meet strict QoS and SLA requirements.

The recently emerged Cloud systems can be used when strict QoS and SLA requirements are applied. In many cases resource requirements are higher than the available Grid infrastructure resources can provide. It is very common that researchers need shorter response time and reliable infrastructure with strict SLAs. Cloud-based infrastructure can fulfill such requirements in many cases; however, the porting costs of complex research application is hard to finance.

Unfortunately, in many cases these pillars are separated from each other and cannot be used simultaneously by the same e-scientist to solve a large-scale single application. Partial results of interconnecting these systems have been achieved in the past. Cluster- and supercomputer-based grids can be referred to as Service Grids (SG) since they provide managed cluster and supercomputer resources as 7/24 services. The OGF PGI (Production Grid Infrastructure) working group has put significant effort to solve the interoperability problem of production service grids, and yet even the interoperation of various cluster-based and supercomputer-based grids is not fully solved. The recently formed EMI (European Middleware Initiative) project aims at integrating the three major European grid middleware systems (ARC, gLite, Unicore) into a unified middleware distribution (UMD). If we could provide seamless interoperability and easy application migration between Grid and

Cloud infrastructure, we can enable the migration of existing scientific applications from Grids towards Clouds, and we can support the better requirements matching between applications and infrastructures. The commercial Cloud infrastructure can support large-scale resource requirements in a reliable way. Solving the interoperability issues between Grids and Clouds, Grid researchers can use in a dynamic way additional Cloud resources if they have exhausted their available Grid resources.

BOINC-based and XtremWeb-based DG systems have been successfully integrated with gLite-based service grids within the EDGeS project [1] at two levels. At the middleware level a bridge technology was developed called as 3G Bridge that enables a seamless extension of gLite-based SGs with DGs, and vice versa gLite-based SGs can support DG systems in a seamless way. At the application level the Grid execution back-end of P-GRADE grid portal was extended with the capability of submitting predefined applications not only into service grids but also to Desktop Grids.

The usage scenario investigated in EDGeS was as follows. Using the P-GRADE grid portal, a user has prepared a large workflow application where some of the workflow nodes require parameter sweep execution with many different parameter sets. In such case the user can direct the execution of such node to a DG system that already supports the application represented by this workflow node. Other nodes of the workflow can be directed to local resources or SG resources as defined by the user. In this way the user can direct different parts of the workflows to the cheapest available Grid resources.

The objective of the research reported in this chapter is to investigate how the application level approach of the EDGeS project can be extended for clouds, i.e., how to direct the execution of the parameter sweep workflow nodes of a P-GRADE workflow not only to DG systems but also to Clouds especially when SLA requirements are more strict than a DG system can satisfy.

The chapter is divided into six sections. After this short introduction, Sect. 6.2 explains the principles of parameter sweep job execution by P-GRADE portal. The next section gives some details on the concept of parameter sweep job submission to various Grids by the 3G Bridge. Section 6.4 introduces several possible solution variants for parameter sweep job submission to Cloud systems. Section 6.5 shows the results of performance measurements. Finally, Sect. 6.6 details some related research results.

## 6.2 Principles of Parameter Sweep Job Submission by P-GRADE Portal

In the academic world science gateways gain more and more popularity especially for developing and running large-scale applications. In Europe one of the most popular generic purpose science gateway is the P-GRADE portal. The basic concepts of the P-GRADE Portal based science gateway solution were mainly developed during the grid era. Therefore P-GRADE portal is used by many national grids (UK NGS,

Grid Ireland, Belgium Grid, SwissGrid, Turkish Grid, Hungarian Grid, Croatian Grid), by several regional grids (South-East European Grid, Baltic Grid, UK White Rose Grid), and by several science specific virtual organizations (Chemistry Grid, Economy Grid, Math Grid, etc.). In recent years P-GRADE portal became popular even outside Europe: in Grid Malaysia established by MIMOS Berhad, Grid Kazakhstan, and Armenian Grid. In total thousands of computers located in more than 45 countries can be accessed via the P-GRADE portal solution.

P-GRADE portal (Parallel Grid Run-time and Application Development Environment) [2, 3] is an open-source tool set consisting of a service-rich, workflow-oriented graphical front-end and a back-end enabling execution in various types of Cluster/Grid/Cloud systems. It supports workflows composed of sequential jobs, parallel jobs, and application services. P-GRADE portal hides the complexity of infrastructure middleware through its high-level graphical web interface, and it can be used to develop, execute, and monitor workflow applications on SG systems (built with Globus, EGEE (LCG or gLite), ARC), on Clusters (using PBS, LSF), and on DG system (using BOINC and XtremWeb middleware). P-GRADE portal installations typically provide the user with access to several middleware technologies, using a single login.

The main features of P-GRADE Portal are:

- Seamless interoperability with the widest range of technologies and Cluster/SG/DG middleware (Globus Toolkit 2, Globus Toolkit 4, LCG, gLite, ARC, BOINC, PBS, LSF, BOINC, XtremWeb) among the available portal/gateway solutions.
- Multigrid access mechanism to simultaneously utilize multiple grid implementations [2].
- Fully compliant with all the security features used in Grids (X.509 certificates, proxy credentials, etc.) and support of other authentication solutions (e.g., Sibboleth).
- Support of workflow-based application design with built-in graphical editor
- Monitoring, accounting and visualization, quota management facilities of infrastructure.
- Legacy application publish and reuse capabilities by the GEMLCA mechanism.
- Large international developer community, world-wide usage with a high number of distributed large-scale, parallel, scientific applications.

The P-GRADE portal applies a DAG (directed acyclic graph) based workflow concept (shown in Fig. 6.1). In a generic workflow, nodes (shown in Fig. 6.1 as large squares) represent jobs, which are basically batch programs to be executed on a computing element. Ports (shown here as small squares around the large ones) represent input/output files the jobs receiving or producing. Arcs between ports represent file transfer operations. The basic semantics of the DAG-based workflow is that a job can be executed if and only if all of its input files are available. This semantics is enforced by the Condor DAGMan workflow manager that is used internally in P-GRADE portal.

In our experience, user communities have shown substantial interest in being able to run programs parallel with different input files. P-GRADE portal supports this

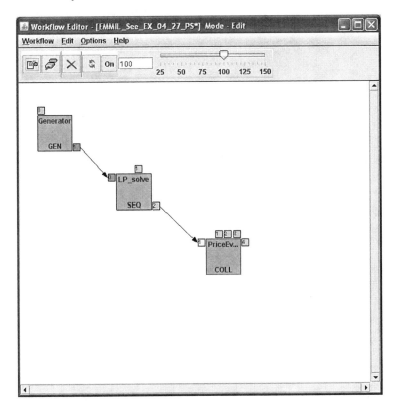

**Fig. 6.1**  Example workflow in P-GRADE Portal

kind of parallelization called Parameter Sweep, or Parameter Study at a high level. The original "job" idea has been extended by two special jobs called Generator and Collector to facilitate the development of parameter sweep (PS) type workflows in P-GRADE portal. The Generator job is used to generate the input files for all parallel jobs automatically (called Automatic Parameter Input Generator) or by a user-uploaded application (called Normal Generator). The Collector will run after all parallel executions are completed and then collects all parallel outputs [4]. All jobs connected to a Generator will run in as many instances as many input files are generated by the Generator (shown in Fig. 6.2).

## 6.3  Principles of Parameter Sweep Job Submission to Various Grids by 3G Bridge

Originally, P-GRADE portal supported parameter sweep job submission to Globus- and gLite-based Service Grids. However, for compute-intensive parameter sweep jobs, Desktop Grids are more ideal than SGs since they are less expensive. Directing

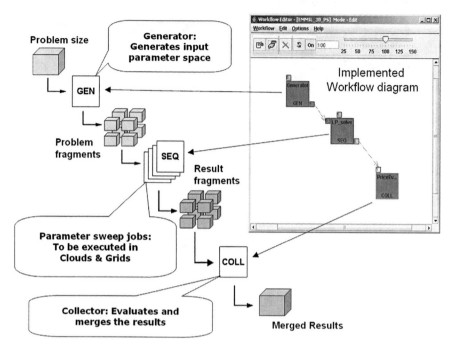

**Fig. 6.2** Parameter sweep solution in P-GRADE Portal with Generator and Collector

these kinds of parameter sweep jobs into DGs will enable one to use the expensive service Grid resources for other type of applications that are not supported by DGs.

In order to direct parameter sweep jobs to Desktop Grids, the 3G Bridge service developed in the EDGeS project has been interfaced with P-GRADE portal. The internal architecture of the 3G Bridge can be seen in Fig. 6.3, where the following components can be identified:

- WSSubmitter provides a web service interface in order to access the 3G Bridge services as a usual web service.
- HTTPD enables to download files from the 3G Bridge machine.
- Job Database is used for storing the jobs to be executed in the target grids.
- Queue Manager is responsible for handling the jobs in the database by using a very simple scheduler for calling the specific plug-ins.
- Grid Handler Interface provides a generic interface above the grid plug-ins.
- Plug-ins responsible for submitting the jobs into different destination infrastructures.
- Download Manager is an internal component of the 3G Bridge that downloads job input files from 3G Bridge clients.

Whenever a new destination infrastructure is to be connected to P-GRADE portal, the corresponding plug-in should be written and attached to the Grid Handler Interface. In the EDGeS project the gLite, XtremWeb, BES, and BOINC plug-ins

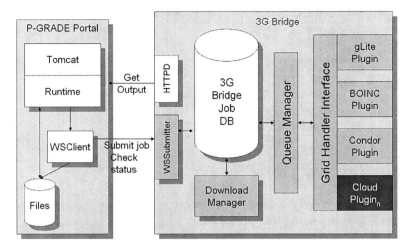

**Fig. 6.3**  Interfacing P-GRADE portal and 3G Bridge

have been developed. With these plug-ins, the user can submit PS jobs to gLite, XtremWeb, ARC and UNICORE, or BOINC Grids.

Once a job should be submitted using the 3G Bridge, P-GRADE Portal runtime makes use of the WSClient application, the client of WSSubmitter. Using this tool, P-GRADE portal can send jobs to the target 3G Bridge service and plug-in. The jobs' input files are sent along with the job submission request, so in this scenario the Download Manager has no real task. Once a job has been submitted, P-GRADE portal periodically updates the job's status using the WSClient application. Finally, if the job has finished, P-GRADE portal fetches the output files from the 3G Bridge service machine using the HTTPD server.

DC-API [5] is used to implement both the BOINC and Condor plug-ins [6]. It provides a simple uniform API for writing master–worker-type applications for different Grids (currently BOINC, Condor, and the Hungarian Cluster Grid are supported). This means that applications using DC-API do not need modifications when moving from one supported platform to another, just relinking with the appropriate DC-API backend library. The Condor and BOINC 3G Bridge plug-ins are basically instances of the same "DC-API-Single" plug-in, but linked with a different DC-API backend library. This also means that the plug-ins are interchangeable; instead of a Condor plug-in, a BOINC one could be used (with some restrictions).

## 6.4  Variants of Creating 3G Bridge Cloud Plug-Ins

In order to submit PS jobs not only to grids but also to Clouds, the possible variants of creating a Cloud plug-in for 3G Bridge should be investigated. Once the Cloud plug-in is available, the portal user can submit PS jobs not only to SGs and DGs but also to Clouds. The Cloud plug-in has to solve three problems to submit and manage PS jobs in Cloud resources:

- Cloud resource management: takes care of allocating Cloud resource for PS jobs when they arrive and removing the Cloud resources when no PS jobs are available for Cloud execution.
- Job submission: Submits PS jobs waiting in the 3G Bridge Job Database to the allocated Cloud resources.
- Job scheduling on Cloud resources: Decides which allocated Cloud resources to be used by the individual PS jobs.

### 6.4.1 The Naive Solution

A naive solution would be that for each incoming PS job, the 3G Bridge plug-in allocates a new Cloud resource and submits the job to this cloud resource. Once the execution of the job on the Cloud resource is finished, the 3G Bridge plug-in removes the cloud resource. This is obviously a very nonoptimal solution that assumes that the number of available Cloud resources is unlimited as well as the budget the user can spend on them. In a realistic scenario the user can afford only a certain number of resources, and once those resources are loaded with jobs, some intelligent job scheduling decision is needed to which already loaded resource to submit the newly arriving PS job. Therefore, more sophisticated solutions have to be investigated where the number of usable Cloud resources has an upper limit. We have considered three basic solution variants:

- Variant 1: Independent Cloud resources with local job managers.
- Variant 2: Communicating Cloud resources with centralized job manager.
- Variant 3: Independent Cloud resources with centralized job managers.

In all three variants we need a job manager that can realize job scheduling on the finite set of resources. Any available job manager system can be considered for this purpose. We selected Condor since it is widely used in the academic community.

### 6.4.2 Independent Cloud Resources with Local Job Managers

The concept of the first variant is shown in Fig. 6.4. The main idea is that on each Cloud resource the 3G Bridge Cloud plug-in deploys a VM image containing a Condor job manager and a Condor worker. The Condor job manager takes care of scheduling the jobs submitted to this resource while the worker will execute them. Due to the Condor job manager, many jobs can be sent to the same Cloud resource. In the 3G Bridge Cloud plug-in, various scheduling algorithms can be applied to select the most optimal Cloud resource. For example, a simple algorithm that realizes both scheduling and Cloud resource management can be the following, where the decision is based on the number of jobs allocated for the Cloud resources:

**Start and submit:**

```
1: if m = 0 ∧ k < n then
2:    instanceid ← start_instance()
3: else
4:    instanceid ← find_min(j₀..jₖ₋₁)
5: end if
6: submit_job(instanceid, jobid)
```

1: **if** $m = 0 \wedge k < n$ **then**
2:     *instanceid* ← *start_instance*()
3: **else**
4:     *instanceid* ← *find_min*($j_0..j_{k-1}$)
5: **end if**
6: submit_job(instanceid, jobid)

**Stop:**

1: **for** $i \leftarrow 0$ to $k - 1$ **do**
2:     **if** $t_i < timestamp(\texttt{"-20 minutes"}) \wedge j_i = 0$ **then**
3:         stop_instance($i$)
4:     **end if**
5: **end for**

**Legend:**

$m$: number of free instances
$k$: total number of running instances
$n$: maximum number of instances
$j_x$: number of jobs queued on instance $x$
$t_x$: timestamp of last execution on instance $x$

According to this algorithm, the 3G Bridge Cloud plug-in is going to start a new instance if there is no free instance and the number of running instances is less than the upper limit of the usable instances, and in this case the job will be submitted to this new instance. Otherwise the new job will be submitted to that instance that has the minimum number of assigned jobs. Notice that this algorithm assumes that the execution times of the PS jobs are approximately equal. It is true for many PS applications, but not always. If this assumption does not hold, the load of the different Cloud resources could significantly vary, and there is no remedy for such problems in this architecture.

The Cloud resource management part of the algorithm will stop an instance if its job queue is empty and there was no activity in the last 20 minutes. This 20 minute buffer time is used to avoid the frequent and useless stopping/restarting activities when the time between the incoming jobs is in the range of several minutes.

The 3G Bridge Cloud plug-in is responsible to allocate the required number of Condor job manager/worker instances in the cloud and submit the PS jobs to the Condor job managers of the instances. The decision to which Condor instance a certain PS job should be sent is also taken by the 3G Bridge Cloud plug-in. It means that all the three functions (cloud resource management, job submission, and job scheduling) should be performed by the 3G Bridge Cloud plug-in. Integrating these three functions into one monolithic plug-in is not a desirable solution since it means that every time a new type of Cloud is to be connected to 3G Bridge, a new complex plug-in with all these functionalities should be developed. A better approach would be to develop an independent simple plug-in for all these three functionalities.

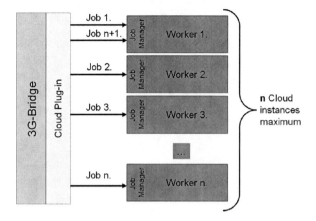

**Fig. 6.4** Variant 1: independent cloud resources with local job managers

As a summary, one can say that the advantages of this variant are as follows:

- Still simple to implement compared to naive solution.
- Uses reliable method for task submission (e.g. Condor WS API).
- The number of used Cloud resources is controllable.

However, there are several serious drawbacks as well:

- A single plug-in manages both the resource allocation and job submission and scheduling that requires major redesign whenever new Clouds (e.g. OpenNebula, Eucalyptus, etc.) or new job managers (PBS, LSF, etc.) are to be used.
- Special prepared VM image is required.
- Since every VM image contains a Job Manager, it raises some overhead. For example, in case of Condor it is around 200 Mb, which has extra cost both in the storage area and in the communication time when the image is moved from the storage to the resources.
- Nodes are separated from each other, and hence there is no way to reschedule jobs from a node to another. As a consequence, some nodes might become over-committed while others are empty.

### 6.4.3 Communicating Cloud Resources with Centralized Job Manager

The architecture and concept of the second variant are shown in Fig. 6.5. The main difference compared with the first variant is that the Condor job manager is placed in a separate image. The other VM images contain only the Condor worker, and the Cloud resources where they are deployed should be able to communicate with the Cloud resource of the Condor job manager.

The 3G Bridge Cloud plug-in becomes significantly simpler since there is no need to schedule the incoming PS jobs among the Condor worker cloud resources. This scheduling will be done by the Condor job manager. The plug-in needs only to start/stop instances for the job manager and submit the jobs to the Condor job manager. The resource management algorithm of the 3G Bridge Cloud plug-in is as follows:

---

**Start and submit:**

1: **if** $tj/k > q \wedge k < n$ **then**
2:     start_worker()
3: **end if**
4: submit_job_to_frontend(jobid)

**Stop:**

1: **for** $i \leftarrow 0$ to $k - 1$ **do**
2:     **if** $tj/(k-1) < q \wedge t_i < timestamp(\texttt{"-20 minutes"})$ **then**
3:         stop_worker($i$)
4:     **end if**
5: **end for**

**Legend:**

  $m$: number of free instances
  $k$: total number of running instances
  $n$: maximum number of instances
  $t$: timestamp of last job submission
  $tj$: total number of jobs in the Cloud
  $q$: preferred maximum job number per worker

---

According to this algorithm, the 3G Bridge cloud plug-in will deploy a new instance only if the average job number per worker surpasses the preferred maximum job number per worker and the number of running instances is less than the per-

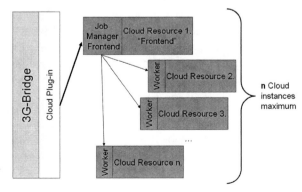

**Fig. 6.5** Variant 2: communicating cloud resources with centralized job manager

mitted maximum number of instances. A worker Cloud resource will be stopped if there was no job submission in the last 20 minutes and the average job number per worker is less than the preferred maximum job number per worker. A worker will be selected based on its "shutdown time window." This can mean different things: Commercial Cloud providers (e.g., Amazon EC2) may charge for instance hours, rounded up, and thus an instance will be charged for an hour regardless if it was running for 15 or 59 minutes. In this case the shutdown time window could be open in the 54–58th minutes of every hour in the lifetime of an instance, and the plug-in is allowed to shutdown the instance during this time window only.

The advantage of this architecture is that the job manager balances tasks between the Cloud resources. As a result, no resource gets overcommitted while others starve. Further advantage is that parallel (MPI, PVM, Map-Reduce, Hadoop) applications can also be executed in this architecture. However, there are several serious drawbacks as well. Two specially prepared VM images are required: one for the workers and one for the Frontend. This solution can be used only in such a Cloud where communication among the Cloud resources is manageable. This architecture also applies a single plug-in although it has to implement only the resource management and job submission functions.

### 6.4.4 Independent Cloud Resources with Centralized Job Managers

The architecture and concept of the third variant are shown in Fig. 6.6. It can be seen that the three functions (resource management, job submission, and job scheduling) of the 3G Bridge cloud plug-in are separated as independent functional units. Even more, the resource management function is realized by two separate components: Cloud Plug-in and Cloud Resource Manager. The latter collects information from the 3G Bridge Condor job queue (Queue 1 in Fig. 6.6) about the number of waiting Condor jobs and from the cloud job queue (in our current implementation, Amazon EC2 or Eucalyptus) about the available Condor worker instances. If the number of worker instances has not reached yet their upper limit and there are waiting Condor jobs in the 3G Bridge Condor job queue, the Cloud Resource Manager sends jobs into the Cloud Plug-in queue (Queue 2 in Fig. 6.6). For every such job, the Cloud Plug-in will deploy a new worker instance in the cloud, and the instances will be kept running as long as the jobs in Queue 2 of the 3G Bridge are in running state. Once an instance is deployed, it will connect to the Condor Submitter/Master unit in order to get a Condor job to execute and will act as an ordinary Condor worker in a Condor pool. The instances are supplied with the address of the Condor Master at startup time, meaning that different instances may connect to different masters and thus to different pools. If there is no waiting Condor job in the 3G Bridge Condor job queue and a certain time has already spent without new arriving Condor job activity, the Cloud Resource Manager cancels some running jobs in the Cloud Plug-in queue, and thus the Cloud instances belonging to those jobs will be terminated by the Cloud plug-in.

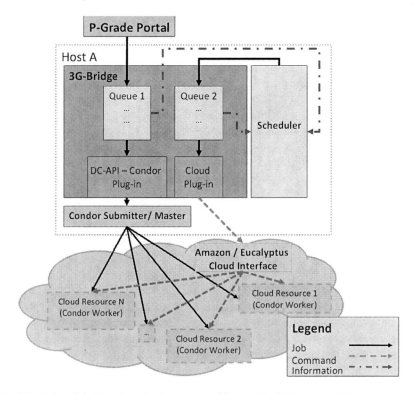

**Fig. 6.6**   Variant 3: independent cloud resources with centralized job manager(s)

PS-jobs arriving at 3G Bridge are submitted to the Condor Submitter/Master unit by the regular DC-API Condor Plug-in. It is the task of the Condor Submitter/Master unit to distribute the Condor jobs among the worker instances running on the cloud resources. Notice that this part of the architecture is exactly the same that is used in any 3G Bridge–Condor interconnection. This is exactly the advantage of this architecture concept that the implementation of the job submission and job scheduling functions does not require any new development.

The architecture of Fig. 6.6 is also very flexible. If someone would like to replace Condor with another job manager, for example, PBS, then it is only the DC-API Condor Plug-in that should be replaced with a PBS Plug-in. The Cloud Plug-in and the Cloud Resource manager still can be used without any modification. Similarly, if someone would like to change Amazon EC2/Eucalyptus for another type of cloud, for example, OpenNebula, then only the cloud interface calls of the Cloud Plug-in should be changed.

If someone would like to use several Clouds simultaneously, the architecture is easily expandable to support it. In this case, an additional cloud Queue and Cloud Plug-in should be added to the architecture, and the Cloud Resource Manager should be extended with information collection capability from the new cloud, too.

## 6.5 Performance Measurements

From the three variants we choose to implement the last one (Variant 3) since it provides simplicity (only workers are running on the Cloud) with enough flexibility (Cloud resources can be started and shutdown independently, and anytime with only minor restrictions, rescheduling of work between nodes is possible). For the development, we used an in-house Eucalyptus 1.51-based local Cloud, which was able to run four instances, each with 512-MB memory. For a real-world deployment, we used Amazon EC2 North-American availability zone with four "High-CPU Medium" (c1.medium) instances to execute 80 jobs. These instances have two virtual cores, so eight cores in total. Each of them has increased CPU performance ("2.5 EC2 Compute Units" each), compared to the default "Small" instance ("1 EC2 Compute Units"), while only costs twice the price thus seem optimal for computation intensive tasks. Although there is a "High-CPU Large" instance with eight cores, that does not provide more performance per CPU or lower cost compared to the medium one, according to the information on the Amazon Web Services Website [7].

Our test was executed with the E-Marketplace Model Integrated Logistics (EMMIL) [8] application, which is solving a multiparameter linear optimization problem to facilitate three-sided negotiation between buyers, sellers, and third-party logistics providers. Its workflow is based on the most supported parallel-execution possibilities of P-GRADE Portal: Parameter Sweep (see Fig. 6.2). 3G Bridge was used for executing the "Parameter sweep jobs" of the EMMIL workflow. We configured EMMIL so that a single PS job would run approximately 4 minutes on a "High-CPU Medium" Amazon EC2 instance.

For the workers, we have developed our own Amazon Machine Image (AMI) similar to [9]. The image is based on Debian Linux 5.0 32-bit and Condor 7.4.2. Each instance is supplied with the IP address of the Host running the 3G Bridge and Condor, and thus they can automatically join the Condor pool (and leave when they are shut down). Figure 6.7 shows our results, namely: (a) Time-lapse of the whole execution, (b) Init phase, and (c) Shutdown phase.

80 jobs were submitted to the 3G Bridge, which then submitted the jobs to the Condor cluster. As Fig. 6.7b shows, 92 seconds were required for the 80 jobs to appear in Condor. Condor caused this, at certain points during the submission the 3G Bridge started to receive "Connection refused" errors when submitting jobs from its queue to Condor. In such case the 3G Bridge enters an exponential fallback sleep cycle between retries, with a maximum (configured) of 32 seconds. It took 159 seconds from the beginning of submission of the jobs until all eight workers joined the pool, but a worker started the first job after 149 seconds. This is the total overhead that includes the submission to 3G Bridge, from 3G Bridge to Condor and the time required for the Cloud resources to join the Condor pool.

The total time for execution of the 80 tasks via 3G Bridge on the Cloud resources took 2860 seconds. Since we did not want to submit more jobs, we issued a manual shutdown for the workers (Fig. 6.7c), by issuing a cancel command for the tasks representing the running Cloud resources in the corresponding 3G Bridge queue.

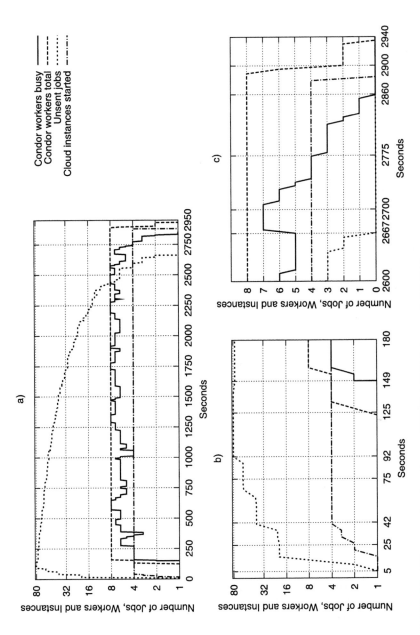

**Fig. 6.7**   EMMIL Jobs (80) executed on four Amazon EC2 High CPU instances via 3G-Bridge

## 6.6 Related Research

Shantenu Jha et al. [10] discuss how Clouds may be treated as higher-level abstraction from Grids. For example, data storage (S3 [11]) and computing (EC2) services offered by Amazon can be thought of as data and storage grids with less exposed semantics (features or options) or can be considered as applications of the underlying service layer. They define the term "usage modes" for different usage patterns or deployment scenarios of applications and "affinity" for describing the level of support of a system for a given usage mode. General-purpose Grids have large semantic feature set, since they try to address a broad variety of usage modes (and applications). They argue that as the semantic complexity decreases, the usability from the end users perspective increases. According to that, Clouds might be "closer" to end users than traditional Grids. This implies that adapting Clouds to Grid interfaces and APIs (e.g., SAGA) first need high-level abstractions for application development and deployment (direct cause of the different "affinities" of the two systems).

Chris Miceli et al. [12] were using "All-pairs" on top of SAGA (Simple API for Grid Applications) based MapReduce [13] implementation. The authors focus on data-intensive tasks, while we focus on computation-intensive tasks, but the general approach and conclusions still match. SAGA is similar to DC-API in means of providing a uniform simple API for GRID applications, but DC-API focuses on master–worker paradigm solely on a few selected platforms rather than aiming for a generic solution. Also SAGA's runtime and (job) adaptors show similarities with the plug-in model of 3G Bridge. The authors executed their experiments on Grids, "Cloud-like infrastructure," and Clouds (Amazon EC2, Eucalyptus, and using Nimbus [14]) with up to 10 workers and with different data-set sizes ranging from 1 GB to 10 GB. On clouds, similar to our approach, they created a custom VM image with SAGA deployed at EC2 but used a bootstrap script that deployed SAGA and dependencies on NIMBUS and Eucalyptus.

Both approaches have their benefits, bootstrapping a "stock" VM image on each boot allows one to use generic images, but an overhead is added every time the VM starts. On the other hand, using custom images allows starting quickly but requires special prepared VM images. The authors claim that EC2 took 90 seconds, while Eucalyptus took 45 seconds to instantiate a VM. It took us 120 seconds on EC2 from issuing a start command for a VM instance until the first Condor worker was ready to accept jobs; on our local Eucalyptus pool this took a lot more time, so we only used Eucalyptus for development and verification. The difference of measurements on EC2 can come from many factors, like the size of the VM image used (it needs to be copied from a store every time a new instance uses it) or like the added overhead of the time required for the worker to join the Condor pool after start. Nevertheless, this shows that there is space for improvement there. The authors also show that there is interoperability between Grids and Clouds using SAGA: the same interface and function calls can be used whether the submission is to Grids or Clouds; this is also true for our approach, since it is only a matter of choosing the appropriate job queue at submission time.

Gideon Juve's "howto" [9] is what we used as a basis for creating the Condor pool of workers in our implementation. This document gives a good starting point and general idea how to connect Condor workers running on Amazon EC2 to an arbitrary Condor pool.

Douglas Thain et al. [15] describe how a Condor-based cluster can act as a cloud computer and propose to extend the Map-Reduce abstraction for cloud computing with new abstraction layers (namely: Map, All-Pairs, Sparse-Pairs, Wavefront, Directed Graph). P-GRADE portal solution is using internally the Condor DAG-Man. Condor-distributed system is well-known and one of the most widely used distributed processing systems, deployed at several thousand institutions around the world, managing several hundred thousand CPU cores in clustered environment.

Mohsen Amini et al. [16] is focusing on so-called marketing-oriented scheduling policies, which can provision extra resources when the local cluster resources are not sufficient to meet the user requirements. Older scheduling policies used in Grids are not working effectively in Cloud environments, mainly because Infrastructure as a Service (IaaS) providers are charging users in a pay-as-you-go manner in an hourly basis for computational resources. To find the trade-off between to buy acquired additional resources from IaaS and reuse existing local infrastructure resources, he proposes two scheduling policies (Cost and Time Optimization scheduling policies) for mixed (commercial and noncommercial) resource environments. The evaluation of the policies was done in Gridbus broker that is originally able to interface with Condor and SGE. However, he is also extending the broker to interact with Amazon EC2. Basically two different approaches were identified on provisioning commercial resources. The first approach is offered by the IaaS providers at resource provisioning level (user/application constraints are neglected: deadline, budget, etc.); the other approach deploys resources focusing at user level (time and/or cost minimization, estimating the workload in advance, etc.).

OpenNebula [17] can hire commercial (Amazon EC2) resources in an on-demand manner, when the local cluster resources are overloaded. Llorente et al. [18] extend OpenNebula to provision additional resources for applications to handle peak load, when the local cluster or grid environment is saturated. Silva et al. [19] proposes a mechanism for creating and deploying optimal amount of resources in Cloud environment, when the workload is hard to predict due to the usage of Bag-of-Task applications.

One solution of application deployment on Cloud infrastructure is detailed in [20], which introduce a way to distribute a parameter-sweep-type application with Aneka (Cloud development and management platform) on Cloud resources. In Aneka enterprise clouds cost-optimization, time-optimization, and conservative time-optimization scheduling algorithms are available.

## 6.7 Conclusion

This chapter introduced the existing connectivity and interoperability issues of Clouds, Grids, and Clusters and showed solutions how such issues can be solved. Firstly we have enumerated the main pillars of e-Science infrastructure ecosystems which are recently extended by Cloud infrastructure. Then the basic principles of parameter sweep job concept and its execution by P-GRADE portal was briefly explained. We have then examined three solution variants in details for processing parameter sweep jobs on Cloud resources. From these three alternatives, the last variant—the most flexible solution using the 3G Bridge developed by the EDGeS project to interconnect different (grid) middleware—was implemented. With performance measurements we have evaluated the implemented solution. Our results show clearly that this solution allows use on-demand cloud resources in a transparent and efficient way with improved scalability. Future work includes performing larger-scale tests and integrating this solution into the EDGeS infrastructure. We also need to handle different requirements of the tasks (e.g., currently we only support 32-bit Linux environment, and no extra dependencies or requirements may be defined for the tasks).

# References

1. Urbah, E., Kacsuk, P., Farkas, Z., Fedak, G., Kecskemeti, G., Lodygensky, O., Marosi, A., Balaton, Z., Caillat, G., Gombas, G., Kornafeld, A., Kovacs, J., He, H., Lovas, R.: EDGeS: bridging EGEE to BOINC and XtremWeb. J. Grid Comput. **7**(3), 335–354 (2009)
2. Kacsuk, P., Sipos, G.: Multi-grid multi-user workflows in the P-GRADE portal. J. Grid Comput. **3**(34), 221–238 (2005)
3. Kacsuk, P., Kiss, T., Sipos, G.: Solving the grid interoperability problem by P-GRADE portal at workflow level. Future Gener. Comput. Syst. **24**(7), 744–751 (2008)
4. Kacsuk, P., Farkas, Z., Herman, G.: Workflow-level parameter study support for production grids. In: International Conference on Computational Science and its Application, ICCSA 2007, Part III, Kuala Lumpur, 2007. Lecture Notes in Computer Science, vol. 2007, pp. 872–885. Springer, Berlin (2007). ISBN:978-3-540-74482-5
5. Marosi, A.C., Gombás, G., Balaton, Z., Kacsuk, P.: Enabling Java applications for BOINC with DC-API. In: Proceedings of the 7th International Conference on Distributed and Parallel Systems, pp. 3–12 (2009)
6. Kacsuk, P., Kovács, J., Farkas, Z., Marosi, A., Gombás, G., Balaton, Z.: SZTAKI desktop grid (SZDG): a flexible and scalable desktop grid system. J. Grid Comput. **7**(4), 439–461 (2009)
7. Amazon EC2 Instance Types. http://aws.amazon.com/ec2/instance-types/. Cited 13 May 2010
8. Kacsuk, P., Kacsukné, B.L., Hermann, G., Balasko, A.: Simulation of EMMIL E-marketplace model in the P-GRADE grid portal. In: Proceedings of ESM'2007 International Conference, Malta, pp. 569–573 (2007)
9. Condor Workers on Amazon EC2: http://www-rcf.usc.edu/juve/condor-ec2/. Cited 13 May 2010
10. Jha, S., Merzky, A., Fox, G.: Using clouds to provide grids with higher levels of abstraction and explicit support for usage modes. Concurr. Comput. Pract. Exp. **21**(8), 1087–1108 (2009)
11. Amazon Simple Storage Service (S3). https://aws.amazon.com/s3. Cited 13 May 2010
12. Miceli, C., Miceli, M., Jha, S., Kaiser, H., Merzky, A.: Programming abstractions for data intensive computing on clouds and grids. In: IEEE International Symposium on Cluster Computing and the Grid, pp. 478–483. IEEE Computer Society, Los Alamitos (2009). ISBN:978-0-7695-3622-4
13. Dean, J., Ghemawat, S.: MapReduce: simplified data processing on large clusters. Commun. ACM **51**(1), 107–113 (2008)
14. NIMBUS. http://www.nimbusproject.org/. Cited 13 May 2010

15. Thain, D., Moretti, C.: Abstractions for Cloud Computing with Condor, Syed Ahson and Mohammad Ilyas, Cloud Computing and Software Services. CRC Press, Boca Raton (2009). ISBN:9781439803158
16. Salehi, M.A., Buyya, R.: Adapting market-oriented scheduling policies for cloud computing. In: 10th International Conference on Algorithms and Architectures for Parallel Processing, ICA3PP 2010, 21–23 May 2010, Busan, Korea
17. Fontán, J., Vázquez, T., Gonzalez, L., Montero, R.S., Llorente, I.M.: OpenNEbula: the open source virtual machine manager for cluster computing. In: Open Source Grid and Cluster Software Conference (2008)
18. Llorente, I.M., Moreno-Vozmediano, R., Montero, R.: Cloud computing for on-demand grid resource provisioning. In: Advances in Parallel Computing (2009)
19. Silva, J.N., Veiga, L., Ferreira, P.: Heuristic for resources allocation on utility computing infrastructures. In: Proceedings of the 6th International Workshop on Middleware for Grid Computing, pp. 9–17 (2008)
20. Cloud based parameter sweeping of a repast simphony simulation model—GUI mode, Technical Note, 24/02/2010. http://community.decisci.com/content/cloud-based-parameter-sweeping-repast-simphony-simulation-model-gui-mode#content. Cited 13 May 2010

# Chapter 7
# Energy Aware Clouds

**Anne-Cécile Orgerie, Marcos Dias de Assunção,
and Laurent Lefèvre**

**Abstract**  Cloud infrastructures are increasingly becoming essential components for providing Internet services. By benefiting from economies of scale, Clouds can efficiently manage and offer a virtually unlimited number of resources and can minimize the costs incurred by organizations when providing Internet services. However, as Cloud providers often rely on large data centres to sustain their business and offer the resources that users need, the energy consumed by Cloud infrastructures has become a key environmental and economical concern. This chapter presents an overview of techniques that can improve the energy efficiency of Cloud infrastructures. We propose a framework termed as Green Open Cloud, which uses energy efficient solutions for virtualized environments; the framework is validated on a reference scenario.

## 7.1 Introduction

Cloud solutions have become essential to current and future Internet architectures as they provide on-demand virtually unlimited numbers of compute, storage and network resources. This elastic characteristic of Clouds allows for the creation of

A.-C. Orgerie (✉)
ENS Lyon, LIP Laboratory (UMR CNRS, INRIA, ENS, UCB), University of Lyon, 46 allée d'Italie, 69364 Lyon Cedex 07, France
e-mail: annececile.orgerie@ens-lyon.fr

M.D. de Assunção · L. Lefèvre
INRIA, LIP Laboratory (UMR CNRS, INRIA, ENS, UCB), University of Lyon, 46 allée d'Italie, 69364 Lyon Cedex 07, France

M.D. de Assunção
e-mail: marcos.dias.de.assuncao@ens-lyon.fr

L. Lefèvre
e-mail: laurent.lefevre@inria.fr

M. Cafaro, G. Aloisio (eds.), *Grids, Clouds and Virtualization*,
Computer Communications and Networks,
DOI 10.1007/978-0-85729-049-6_7, © Springer-Verlag London Limited 2011

computing environments that scale up and down according to the requirements of distributed applications.

Through economies of scale, Clouds can efficiently manage large sets of resources; a factor that can minimize the cost incurred by organizations when providing Internet services. However, as Cloud providers often rely on large data centres to sustain their business and supply users with the resources they need, the energy consumed by Cloud infrastructures has become a key environmental and economical concern. Data centres built to support the Cloud computing model can usually rely on environmentally unfriendly sources of energy such as fossil fuels [2].

Clouds can be made more energy efficient through techniques such as resource virtualization and workload consolidation. After providing an overview of energy-aware solutions for Clouds, this chapter presents an analysis of the cost of virtualization solutions in terms of energy consumption. It also explores the benefits and drawbacks that Clouds could face by deploying advanced functionalities such as Virtual Machine (VM) live migration, CPU throttling and virtual CPU pinning. This analysis is important for users and administrators who want to devise resource allocation schemes that endeavor to reduce the $CO_2$ footprint of Cloud infrastructures.

As we attempt to use recent technologies (i.e. VM live migration, CPU capping and pinning) to improve the energy efficiency of Clouds, our work can be positioned in the context of future Clouds. This work proposes an energy-aware framework termed as Green Open Cloud (GOC) [21] to manage Cloud resources. Benefiting from the workload consolidation [34] enabled by resource virtualization, the goal of this framework is to curb the energy consumption of Clouds without sacrificing the quality of service (in terms of performance, responsiveness and availability) of user applications. All components of the GOC architecture are presented and discussed: Green policies, prediction solutions and network presence support. We demonstrate that under a typical virtualized scenario, GOC can reduce the energy used by Clouds by up to 25% compared to basic Cloud resource management.

The remaining part of this chapter is organized as follows. Section 7.2 discusses energy-aware solutions that can be applied to Clouds. Then, the energy cost of virtual machines is investigated in Sect. 7.3. The GOC architecture is described in Sect. 7.4 and evaluated on a typical virtualized scenario Sect. 7.5.

## 7.2 Overview of Energy Aware Techniques for Clouds

Current Web applications demand highly flexible hosting and resource provisioning solutions [35]. The rising popularity of social network Web sites, and the desire of current Internet users to store and share increasing amounts of information (e.g. pictures, movies, life-stories, virtual farms) have required scalable infrastructure. Benefiting from economies of scale and recent developments in Web technologies, data centres have emerged as a key model to provision resources to Web applications and deal with their availability and performance requirements. However, data centres are often provisioned to handle sporadic peak loads, which can result in low resource utilization [15] and wastage of energy [12].

The ever-increasing demand for cloud-based services does raise the alarming concern of data centre energy consumption. Recent reports [29] indicate that energy costs are becoming dominant in the Total Cost of Ownership (TCO). In 2006, data centres represented 1.5 percent of the total US electricity consumption. By 2011, the current data centre energy consumption could double [31] leading to more carbon emissions. Electricity becomes the new limiting factor for deploying data centre equipments.

A range of technologies can be utilized to make cloud computing infrastructures more energy efficient, including better cooling technologies, temperature-aware scheduling [9, 24, 30], Dynamic Voltage and Frequency Scaling (DVFS) [14, 33], and resource virtualization [36]. The use of VMs [3] brings several benefits including environment and performance isolations; improved resource utilization by enabling workload consolidation; and resource provisioning on demand. Nevertheless, such technologies should be analysed and used carefully for really improving the energy-efficiency of computing infrastructures [23]. Consolidation algorithms have to deal with the relationship between performance, resource utilization and energy, and can take advantage from resource heterogeneity and application affinities [34]. Additionally, techniques such as VM live-migration [6, 13] can greatly improve the capacities of Cloud environments by facilitating fault management, load balancing and lowering system maintenance costs. VM migration provides a more flexible and adaptable resource management and offers a new stage of virtualization by removing the concept of locality in virtualized environments [38].

The overhead posed by VM technologies has decreased over the years, which has expanded their appeal for running high-performance computing applications [37] and turned virtualization into a mainstream technology for managing and providing resources for a wide user community with heterogeneous software-stack requirements. VM-based resource management systems, such as Eucalyptus [26] and OpenNebula [10], allow users to instantiate and customize clusters of virtual machines atop the underlying hardware infrastructure. When applied in a data centre environment, virtualization can allow for impressive workload consolidation. For instance, as Web applications usually present variable user population and time-variant workloads, virtualization can be employed to reduce the energy consumed by the data centre environment through server consolidation, whereby VMs running different workloads can share the same physical host. By consolidating the workload of user applications into fewer machines, unused servers can potentially be switched off or put in low energy consumption modes. Yet attracting virtualization is, its sole use does not guarantee reductions in energy consumption. Improving the energy efficiency of Cloud environments with the aid of virtualization generally calls for devising mechanisms that adaptively provision applications with resources that match their workload demands and utilizes other power management technologies such as CPU throttling and dynamic reconfiguration, allowing unused resources to be freed or switched off.

Existing work has proposed architectures that benefit from virtualization for making data centres and Clouds more energy efficient. The problem of energy-efficient resource provisioning is commonly divided into two subproblems [22]: at micro- or host level, power management techniques are applied to minimize the number of resources used by applications and hence reduce the energy consumed by an individual host; and at a macro-level, generally a Resource Management System (RMS) strives to enforce scheduling and workload consolidation policies that attempt to reduce the number of nodes required to handle the workloads of user applications or place applications in areas of a data centre that would improve the effectiveness of the cooling system. Some of the techniques and information commonly investigated and applied at a macro- or RMS-level to achieve workload consolidation and energy-efficient scheduling include:

- Applications workload estimation;
- The cost of adaptation actions;
- Relocation and live-migration of virtual machines;
- Information about server-racks, their configurations, energy consumption and thermal states;
- Heat management or temperature-aware workload placement aiming for heat distribution and cooling efficiency;
- Study of application dependencies and creation of performance models; and
- Load balancing amongst computing sites;

Server consolidation has been investigated in previous work [5, 8, 16, 18, 20, 34, 39, 40]. A key component of these systems is the ability to monitor and estimate the workload of applications or the arrival of user requests. Several techniques have been applied to estimate the load of a system, such as exponential moving averages [4], Kalman filters [17], autoregressive models, and combinations of methods [5, 19]. Provisioning VMs in an IaaS environment poses additional challenges as information about the user applications is not always readily available. Section 7.4.3 describes an algorithm for predicting the characteristics of advance reservation requests that resemble requests for allocating virtual machines.

Fitted with workload-estimation techniques, these systems provide schemes to minimize the energy consumed by the underlying infrastructure while minimizing costs and violations of Service Level Agreements (SLAs). Chase et al. [5] introduced MUSE, an economy-based system that allocates resources of hosting centres to services aiming to minimize energy consumption. Services bid for resources as a function of delivered performance whilst MUSE switches unused servers off. Kalyvianaki et al. [18] introduced autonomic resource provisioning using Kalman filters. Kusic et al. proposed a lookahead control scheme for constantly optimizing the power efficiency of a virtualized environment [20]. With the goal of maximizing the profit yielded by the system while minimizing the power consumption and SLA violations, the provisioning problem is modelled as a sequential optimization under uncertainty and is solved using the lookahead control scheme. Placement of applications and scheduling can also take into account the thermal states or the heat dissipation in a data centre [24]. The goal is scheduling workloads in a data centre and the heat they generate, in a manner that minimises the energy required by the cooling infrastructure, hence aiming to minimize costs and increase the overall reliability of the platform.

Although consolidation fitted with load forecasting schemes can reduce the overall number of resources used to serve user applications, the actions performed by RMSs to adapt the environment to match the application demands can require the relocation and reconfiguration of VMs. That can impact the response time of applications, consequently degrading the QoS perceived by end users. Hence, it is important to consider the costs and benefits of the adaptation actions [40]. For example, Gueyoung et al. [16] have explored a cost-sensitive adaptation engine that weights the potential benefits of reconfiguration and their costs. A cost model for each application is built offline, and to decide when and how to reconfigure the VMs, the adaptation engine estimates the cost of adaptation actions in terms of changes in the utility, which is a function of the application response time. The benefit of an action is given by the improvement in application response time and the period over which the system remains in the new configuration.

Moreover, consolidation raises the issue of dealing with necessary redundancy and placement geodiversity at the same time. Cloud providers, as Salesforce.com for example, that offer to host entire websites of private companies [11], do not want to lose entire company websites because of power outages or network access failures. Hence, outage and blackout situations should be anticipated and taken into account in the resource management policies [32].

While the macro-level resource management performs actions that generally take into account the power consumption of a group of resources or the whole data centre, at the host-level the power management is performed by configuring parameters of the hypervisor's scheduler, such as throttling of Virtual CPUs (VCPU), and using other OS specific policies. In the proposed architectures, hosts generally run a local resource manager that is responsible for monitoring the power consumption of the host and optimizing it according to local policies. The power management capabilities available in virtualized hosts has been categorized as [25]: "soft" actions such as CPU idling and throttling; "hard" actions like DVFS; and consolidating in the hypervisor. CPU idling or soft states consist in changing resource allotments of VMs and attributes of the hypervisor's scheduler (e.g. number of credits in Xen's credit scheduler) to reduce the CPU time allocated to a VM so that it consumes less power. Hard actions comprise techniques such as scaling the voltage and frequency of CPUs. Consolidation can also be performed at the host-level where the VCPUs allocated to VMs can be configured to share CPU cores, putting unused cores in idle state, hence saving the energy that would otherwise be used by the additional core to run a VM.

Nathuji and Schwan [25] presented VirtualPower, a power management system for virtualized environments that explores both hardware power scaling and software-based methods to control the power consumption of underlying platforms. VirtualPower exports a set of power states to VM guests that allow guests to use and act upon these states, thus performing their own power management policies. The soft states are intercepted by Xen hypervisor and are mapped to changes in the underlying hardware such as CPU frequency scaling according to the virtual power management rules. The power management policies implemented in the guest VMs are used as "hints" by the hypervisor rather than executable commands. They also evaluate the power drawn by cores at different frequency/voltage levels and suggest that such technique be used along with soft schemes.

## 7.3 Investigating the Energy Consumption of Virtual Machines

With the aim of integrating soft schemes such as CPU throttling, the next sections describe simple experiments to evaluate the distance between the idle consumption and the consumption at high utilization of virtualized servers. The experiments evaluate the additional power drawn by VMs and the impact of operations such as CPU idling, consolidation at the host level and VM live-migration.

### 7.3.1 Experimental Scenario

The evaluation has been performed on a testbed composed of HP Proliant 85 G2 servers (2.2 GHz, 2 duo core CPUs per node) with Xen open source 3.4.1. Each node is connected to an external wattmeter that logs its instant power consumption; one measurement is taken each second. The storage of these energy logs is performed by a data collector machine. The precision of measurements of this setup is 0.125 watts, whereas the frequency of measurements is one second.

### 7.3.2 Virtual Machine Cost

In order to be energy efficient, virtual machines should ideally start and halt quickly and without incurring too much energy usage. It is hence crucial to understand the cost of basic virtual machine operations and statuses (such as boot, run, idle, halt) in terms of energy consumption.

The experiments reported here describe the stages of booting, running and shutting down virtual machines. Each virtual machine is configured to use one VCPU without pinning (i.e. we do not restrict which CPUs a particular VCPU may run on using the generic *vcpu-pin* interface) and no capping of CPU credits. We measure the energy consumption by running different configurations, one at a time, each with a different number of virtual machines, from one to six virtual machines on one physical resource.

The graph in Fig. 7.1 shows the energy consumption of virtual machines that are initialized but do not have a workload (i.e. they do not execute any application after booting). As shown by this figure, the start and shutdown phases of VMs consume an amount of energy that should not be ignored when designing resource provisioning or consolidation mechanisms. The graph in Fig. 7.2, on the other hand, shows the consumption of virtual machines that execute a CPU intensive sample application

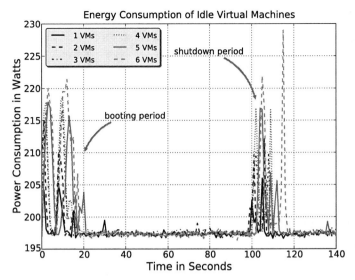

**Fig. 7.1** Energy consumption of different numbers of idle virtual machines on one physical machine

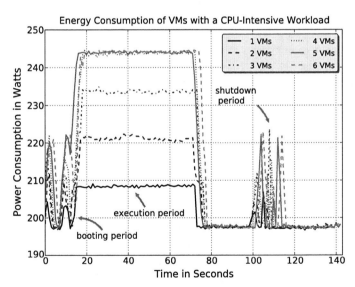

**Fig. 7.2** Energy consumption of virtual machines running a CPU intensive workload on a shared physical resource

(i.e. cpuburn[1]) once they finish booting. The CPU intensive workload runs for 60

---

[1] cpuburn is a test application that attempts to use 100% of CPU.

seconds once the virtual machine is initialized, and we wait for some time before shutting it down to recognize more clearly the shutdown stage in the graphs.

As shown by Fig. 7.2, the increase in energy consumption resulting from increasing the number of virtual machines in the system depends largely on the number of cores in use. Although the 5th and 6th VMs seem to come at zero cost in terms of energy consumption since all the cores were already in use, in reality the application performance can be degraded. We have not evaluated the performance in this work since we run a sample application, but in future we intend to explore the performance degradation and the trade-off between application performance and energy savings.

### 7.3.3 Migration Cost

This experiment evaluates the energy consumption when performing live migration of four virtual machines running a CPU intensive workload. After all virtual machines are initialized, at time 40 seconds, the live migration starts, and one virtual machine is migrated at every 30 seconds. The results are summarized in Fig. 7.3.

The figure shows that migrating VMs lead to an increase in energy consumption during the migration time, the factor evidenced by the asymmetry between the lines. Even though the VMs migrated in this scenario had only 512 MB of RAM, the experiment demonstrates that when migrating with the goal of shutting down unused resources, one must consider that during the migration, two machines will be consuming energy.

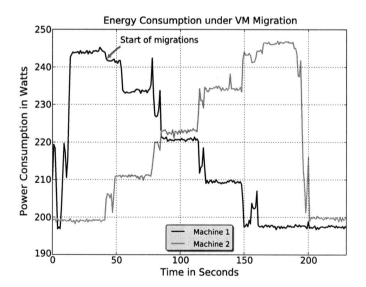

**Fig. 7.3** Migration of four virtual machines starting after 40 seconds, one migration every 30 seconds

### 7.3.4 Capping and Pinning of VCPUs

This experiment measures the energy consumption of virtual machines when we vary parameters of the credit scheduler used by Xen hypervisor. The first parameter evaluated is the cap of CPU utilization. Valid values for the cap parameter when using one virtual CPU are between 1 and 100, and the parameter is set using the *xm sched-credit* command. Initially, we run a virtual machine with a CPU-intensive workload and measure the energy consumption as we change the CPU cap. The experiment considers the following caps: 100, 80, 60, 40 and 20. Once the virtual machine is initialized and the CPU intensive workload is started, the virtual machine remains during one minute under each cap and is later shut down. The results are summarized in Fig. 7.4.

Although capping has an impact on power consumption, this impact is small when considering only one core, and under some cap values, the energy consumption does not remain stable. However, the difference in consumption is more noticeable when throttling the CPU usage of multiple VMs.

When virtual machines are initialized, the credit scheduler provided by Xen hypervisor is responsible for deciding which physical cores the virtual CPUs will utilize. However, the assignment of virtual CPUs to physical cores can be modified by the administrator via Xen's API by restricting the physical cores used by a virtual machine, operation known as "pinning". The second set of experiments described here evaluate the energy consumption when changing the default pinning carried out by Xen's hypervisor.

This experiment first initializes two VMs with the CPU intensive workload; then, leaves them run under default pinning for one minute; after that, throttles the VCPUs

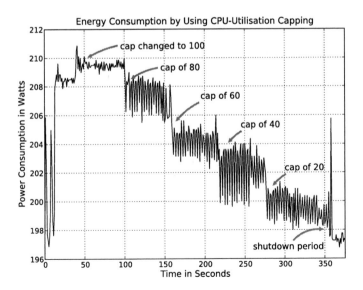

**Fig. 7.4** Energy consumption of a virtual machine under different CPU utilization caps

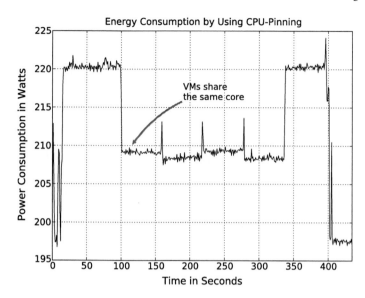

**Fig. 7.5** Energy consumption by making the virtual CPUs of two VMs share the same core

by forcing them to share the same core; next, changes the core to which the VCPUs are pinned at each minute; after that, removes pin restrictions; and finally shuts down the VMs. Figure 7.5 summarizes the results.

The energy consumption reduces as the number of utilized cores in the system decreases. Although this may look obvious, the results demonstrate that it is possible to consolidate a workload at the host level and minimize the energy consumption by using capping. For example, a hosting centre may opt for consolidating workloads if a power budget has been reached or to minimize the heat produced by a group of servers. Our future work will be investigating the trade-off's between performance degradation and energy savings achieved by VCPU throttling when provisioning resources to multi-tier Web applications.

## 7.4 The Green Open Cloud

### 7.4.1 The Green Open Cloud Architecture

The energy footprint of current data centres has become a critical issue, and it has been shown that servers consume a great amount of electrical power even when they are idle [28]. Existing Cloud architectures do not take full advantage of recent techniques for power management, such as Virtual Machine (VM) live migration, advance reservations, CPU idling, and CPU throttling [25].

We attempt to make use of recent technologies to improve the energy efficiency of Clouds, thus placing our work in the context of future Clouds. This work proposes

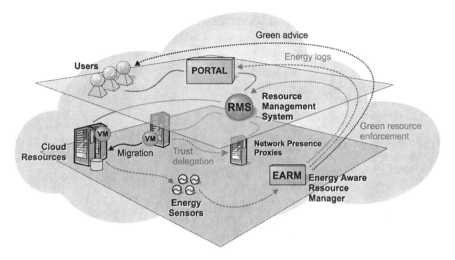

**Fig. 7.6**  The GOC architecture

an energy-aware framework termed as Green Open Cloud (GOC) [21] to manage Cloud resources. Benefiting from the workload consolidation [34] enabled by resource virtualization, the goal of this framework is to curb the energy consumption of Clouds without sacrificing the quality of service (in terms of performance, responsiveness and availability) of user applications.

In the proposed architecture, users submit their reservation requests via a Cloud portal (e.g. a Web server) (Fig. 7.6). A reservation request submitted by a user to the portal contains the required number of VMs, its duration and the wished start time. GOC manages an agenda with resource reservations; all reservations are strict, and once approved by GOC, their start times cannot change. This is like a green Service Level Agreement (SLA) where the Cloud provider commits to give access to required resources (VMs) during the entire period that is booked in the agenda. This planning provides a great flexibility to the provider that can have a better control on how the resources are provisioned.

Following the pay-as-you-go philosophy of Clouds, the GOC framework delivers resources in a pay-as-you-use manner to the provider: only utilized resources are powered on and consume electricity. The first step to reduce electricity wastage is to shut down physical nodes during idle periods. However, this approach is not trivial as a simple on/off policy can be more energy consuming than doing nothing because powering nodes off and on again consumes electricity and takes time. Hence, GOC uses prediction algorithms to avoid frequent on/off cycles. VM migration is used to consolidate the load into fewer resources, thus allowing the remaining resources to be switch off. VCPU pinning and capping are used as well for workload consolidation. GOC can also consolidate workloads considering time by aggregating resource reservations. When a user submits a reservation request, the Green Open Cloud framework suggests alternative start times for the reservation request along with the predicted amount of energy it will consume under the different scenarios. In this way, the user can make an informed choice and favor workload aggregation in time, hence avoiding excessive on/off cycles. On/off algorithms also pose problems for resource management systems, which can interpret a switched-off node as a failure. To address this issue, GOC uses a trusted proxy that ensures the nodes' network presence; the Cloud RMS communicates with this proxy instead of the switched-off nodes (Fig. 7.6).

The key functionalities of the GOC framework are to:

- monitor Cloud resources with energy sensors to take efficient management decisions;
- provide energy usage information to users;
- switch off unused resources to save energy;
- use a proxy to ensure network presence of switched-off resources and thus provide inter-operability with different RMSs;
- use VM migration to consolidate the load of applications in fewer resources;
- predict the resource usage to ensure responsiveness; and
- give "green" advice to users in order to aggregate resource requests.

The GOC framework works as an overlay atop existing Cloud resource managers. GOC can be used with all types of resource managers without impacting on their workings, such as their scheduling policies. This modular architecture allows a great flexibility and adaptivity to any type of future Cloud's RMS architecture. The GOC framework relies on energy sensors (wattmeters) to monitor the electricity consumed by the Cloud resources. These sensors provide direct and accurate assessment of GOC policies, which helps it in using the appropriate power management solutions.

The GOC architecture, as described in Fig. 7.6, comprises:

- a set of energy sensors providing dynamic and precise measurements of power consumption;
- an energy data collector which stores and provides the energy logs through the Cloud Web portal (to increase the energy-awareness of users);
- a trusted proxy for supporting the network presence of switched-off Cloud resources; and
- an energy-aware resource manager and scheduler which applies the green policies and gives green advice to users.

The Cloud RMS manages the requests in coordination with the energy-aware resource manager which is permanently linked to the energy sensors. The energy-aware resource manager of GOC can be implemented either as an overlay on the existing Cloud RMS or as a separate service. Figure 7.7 depicts a scenario where GOC is implemented as an overlay on the existing Cloud RMS. In the resource manager, the white boxes represent the usual components of a future Cloud RMS (with agenda), whereas the shaded boxes depict the GOC functionalities. These add-ons are connected to the RMS modules and have access to the data provided by users (i.e. submitted requests) and the data in the reservation agenda.

When a user submits a request, the *admission control* module checks whether it can be admitted into the system by attempting to answer questions such as: Is the

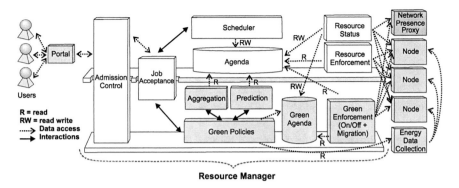

**Fig. 7.7** The GOC resource manager

user allowed to use this Cloud? Is the request valid? Is the request compliant with the Cloud's usage chart? Then, the *job acceptance* module transfers the request to the *scheduler* and to the *green resource manager* (also called energy-aware manager and labeled "*Green Policies*" in Fig. 7.7). The *scheduler* queries the *agenda* to see whether the requested reservation can be placed into it whilst the *green resource manager* uses its *aggregation* and *prediction* modules to find other less energy-consuming possibilities to schedule this reservation in accordance with its green policies. All the possible schedules (from the Cloud's *scheduler* and from the *green resource manager*) are then transferred to the *job acceptance* module which presents them to the user and prompts her for a reply.

The *agenda* is a database containing all the future reservations and the recent history (used by the prediction module). The *green agenda* contains all the decisions of the green resource manager: on/off and VM migrations (when to switch on and off the nodes and when to migrate VMs). These decisions are applied by the *green enforcement* module. The *resource enforcement* module manages the reservations (gives the VMs to the users) in accordance with the *agenda*. The *resource status* component periodically polls the nodes to know whether they have hardware failures. If the nodes are off, the component queries the *network presence proxy*, so the nodes are not woken up unnecessarily. The *energy data collector* module monitors the energy usage of the nodes and gives access to this information to the *green resource manager*. These data are also put on the Cloud Web portal, so users can see their energy impact on the nodes and increase their energy-awareness.

## 7.4.2 Network Presence

As we switch off unused nodes, they do not reply to queries made by the Cloud resource manager, which can be considered by the RMS as a resource failure. This problem is solved by using a trusted proxy to ensure the network presence of the Cloud resources. When a Cloud node is switched off, all of its basic services (such as ping or heartbeat services) are migrated to the proxy that will then answer on behalf of the node when asked by the resource manager. The key issue is to ensure the security of the infrastructure and to avoid the intrusion of malicious nodes. Our trust-delegation model is described in more details in previous work [7]. This mechanism allows a great adaptivity of the GOC framework to any Cloud RMS.

### 7.4.3 Prediction Algorithms

As reservations finish, the prediction algorithm predicts the next reservation for the freed resources. If the predicted reservation of a resource is too close (in less than $T_s$ seconds), the resource remains powered on (it would consume more energy to switch it off and on again), otherwise it is switched off. This idea is illustrated by Fig. 7.8.

For a given resource, $T_s$ is given by the following formula:

$$T_s = \frac{E_{\text{ON}\to\text{OFF}} + E_{\text{OFF}\to\text{ON}} - P_{\text{OFF}}(\delta_{\text{ON}\to\text{OFF}} + \delta_{\text{OFF}\to\text{ON}})}{P_{\text{idle}} - P_{\text{OFF}}}$$

where $P_{\text{idle}}$ is the idle consumption of the resource (in Watts), $P_{\text{OFF}}$ is the power consumption when the resource is off (in watts), $\delta_{\text{ON}\to\text{OFF}}$ is the duration of the resource shutdown (in seconds), $\delta_{\text{OFF}\to\text{ON}}$ is the duration of the resource boot, $E_{\text{ON}\to\text{OFF}}$ is the energy consumed to switch off the resource (in Joules) and $E_{\text{OFF}\to\text{ON}}$ is the energy consumed to switch on the resource (in Joules).

In the same way, we define temporal parameters based on the energy consumption of the resources to know in which case it would be more energy efficient to

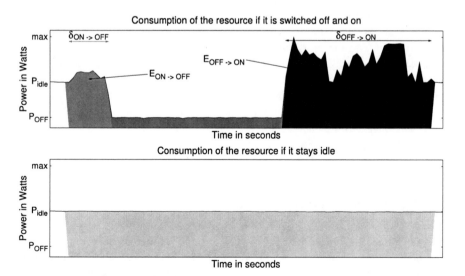

**Fig. 7.8** Definition of $T_s$

use VM migration. $T_m$ is the bound on the remaining reservation's duration beyond which migration is more energy efficient: if the reservation is still running for more than $T - m$ seconds, the VMs should be migrated, otherwise they stay on their respective physical hosts. As the migration takes time, migration of small jobs is not useful (with a duration lower than 20 seconds in that example). This lower time bound is denoted $T_a$. If a migration is allowed, the *green enforcement* module also waits $T_a$ seconds after the beginning of a job before migrating it. It indeed allows initializing the job and the VM, so the migration is simplified, and we avoid the migration of small jobs, crashed jobs or VMs (in case of technical failure or bad configuration of the VM).

For the next reservation, the predicted arrival time is the average of the inter-submission time of the previous jobs plus a feedback. The feedback is computed with the previous predictions; it represents an average of the errors made by computing the $n$ previous predictions ($n$ is an integer). The error is the difference between the true observation and the predicted value. For estimating other features of the reservation (size in number of resources and length), the same kind of algorithm is used (average of the previous values at a given time).

We have seen in previous work [27] that even with a small $n$ (5 for example) we can obtain good results (70% of good predictions on experimental Grid traces). This prediction model is simple, but it does not need many accesses and is really fast to compute, which are crucial features for real-time infrastructures. This prediction algorithm has several advantages: it works well, it requires a small part of history (no need to store big amounts of data), and it is fast to compute. Therefore, it will not delay the whole reservation process, and as a small part of the history is used, it is really responsive and well adapted to request bursts as it often occurs in such environments.

### 7.4.4 Green Policies

Among the components of a Cloud architecture, up to now, we have focused on virtualization, which appears as the main technology used in these architectures. It also uses migration to dynamically unbalance the load between the Cloud nodes in order to shut down some nodes and thus save energy.

However, GOC algorithms can employ other Cloud components, like accounting, pricing, admission control and scheduling. The role of the green policies is to implement such strong administrator decisions at the Cloud level. For example, the administrator can establish a power budget per day or per user. The *green policies* module will reject any reservation request that exceeds the power budget limit. This mechanism is similar to a green SLA between the user and the provider.

Green accounting will give to "green" users (the users that accept to delay their reservation in order to aggregate it with others to save energy) credits which are used to give more priority to the requests of these users when a burst of reservation requests arrives. A business model can also lean on this accounting to encourage users to be energy aware.

## 7.5  Scenario and Experimental Results

### 7.5.1  Experimental Scenario

This section describes the first results obtained with a prototype of GOC. Our experimental platform consists of HP Proliant 85 G2 Servers (2.2 GHz, 2 dual core CPUs per node) with XenServer 5.0[2] [1, 3] on each node.

The following experiments aim to illustrate the working of GOC infrastructure in order to highlight and to compare the energy consumption induced by a Cloud infrastructure with different management schemes. Our experimental platform consists of two identical Cloud nodes, one resource manager that is also the scheduler and one energy data collector. All these machines are connected to the same Ethernet router. In the following we will call 'job' a reservation made by a user to have the resources at the earliest possible time. When a user submits a reservation, she specifies the length in time and the number of resources required. Although this reservation mechanism may look unusual, it is likely to be used by next-generation Clouds where frameworks would support these features to help cloud providers avoid over-provisioning resources. Having scheduling reservations with limits, such as a time duration, will help Cloud managers to manage their resources in a more energy-efficient way. This is, for instance, the case of a user with budget constraints and whose reservations will have a defined time-frame that reflects how much the user is willing to pay for using the resources. In addition, this reflects a scenario where service clouds with long-live applications have their resource allotments adapted according to changes in cost conditions (e.g. cost changes in electricity) and scheduled maintenances.

Our job arrival scenario is as follows:

- $t = 10$: 3 jobs of length equals to 120 seconds each and 3 jobs of length 20 seconds each;
- $t = 130$: 1 job of length 180 seconds;
- $t = 310$: 8 jobs of length 60 seconds each;
- $t = 370$: 5 jobs of length 120 seconds each, 3 jobs of length 20 seconds each and 1 job of length 120 seconds, in that order.

The reservations' length is short in order to keep the experiment and the graph representation readable and understandable. Each reservation is a computing job with a *cpuburn* running on the VM. We have seen that the boot and the shutdown of a VM are less consuming than a *cpuburn* and that an idle VM does not consume any noticeable amount of energy (see Fig. 7.2). Hence, we will just include the *cpuburn* burn phase in the graphs in order to reduce their length and improve their readability (VMs are booted before the experiment start and halted after the end of the experiment).

---

[2]XenServer is a cloud-proven virtualization platform that delivers the critical features of live migration and centralized multi-server management.

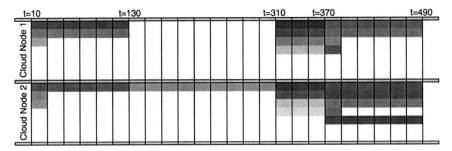

**Fig. 7.9** Gantt chart for the round-robin scheduling

These experiments, although in a small scale for clarity sake, represent the most unfavorable case in terms of energy consumption as *cpuburn* fully uses the CPU which is the most energy consuming component of physical nodes. Moreover, as CPU is a great energy consuming component, *cpuburn* jobs are accurately visible on the power consumption curves: clear steps are noticeable for each added VM (as previously noticed in Fig. 7.2).

Each node can host up to seven VMs. All the hosted VMs are identical in terms of memory and CPU configuration, and they all host a *Debian etch* distribution. As our infrastructure does not depend on any particular resource manager, we do not change the scheduling of the reservations and the assignment of the virtual machines to physical machines. The only exception is when a physical node is off, where we then attribute its jobs to the awoken nodes if they can afford it; otherwise we switch it on.

In order to validate our framework and to prove that it achieves energy saving regardless of the underlying scheduler, we have studied two different schedulings:

- *round-robin*: first job is assigned to the first Cloud node, the second job to the second node, and so on. When all the nodes are idle, the scheduler changes their order (we do not always attribute the first job in the queue to the first node). The behavior of this scheduling mechanism with the previously defined job arrival scenario is shown in Fig. 7.9. This is a typical distributed-system scheduling algorithm.
- *unbalanced*: the scheduler puts as many jobs as possible on the first Cloud node, and, if there are still jobs left in the queue, it uses the second node, and so on (as before, when all the nodes are idle, we change the order to balance the roles). We can see this scheduling with the previously defined job arrival scenario in Fig. 7.10. This scheduling is broadly used for load consolidation.

These two scheduling algorithms are well known and widely used in large-scale distributed system schedulers. For each of these algorithms, we use four scenarios to see the difference between our VM management scheme and a basic one. The four scenarios are as follows:

- *basic*: no changes are made. This scenario represents the Clouds with no power management.

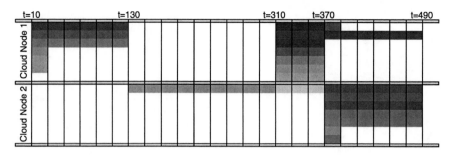

**Fig. 7.10** Gantt chart for the unbalanced scheduling

- *balancing*: migration is used to balance the load between the Cloud nodes. This scenario presents the case of a typical load balancer which aims to limit node failures and heat production.
- *on/off*: unused nodes are switched off. This is the scenario with a basic power management;
- *green*: we switch off the unused nodes, and we use migration to unbalance the load between Cloud nodes. This allows us to aggregate the load on some nodes and switch off the other ones. This is the scenario that corresponds to GOC.

### 7.5.2 Results

For each scheduling, we have run the four scenarios, one at a time, on our experimental platform, and we have logged the energy consumption (one measure per node and per second at the precision of 0.125 watts). A more extensive discussion on the results is available in previous work [21].

Figure 7.11 shows the course of experiment for the two nodes with the round-robin scheduling applied to the green scenario. The upper part of the figure shows the energy consumption of the two nodes during the experiment, while the lower part presents the Gantt chart (time is in seconds).

At time $t = 30$, the second job (first VM on Cloud node 2) is migrated to free Cloud node 1 and then to switch it off. This migration does not occur before $T_a = 20$ seconds. Some small power peaks are noticeable during the migration. This confirms the results of Sect. 7.3.3 stating that migration is not really costly if it is not too long.

This migration leads to the reallocation of the job starting at $t = 130$ on Cloud node 1 since Cloud node 2 has been switched off and Cloud node 1 is available. At $t = 200$, Cloud node 2 is booted to be ready to receive the jobs starting at $t = 310$. During the boot, an impressive consumption peak occurs. It corresponds to the physical start of all the fans and the initialization of all the node components. Yet, this peak is more than compensated by the halting of the node ($P_{OFF} = 20$ watts) during the time period just before the boot since the node inactivity time was greater than $T_s$ (defined in Sect. 7.4.3).

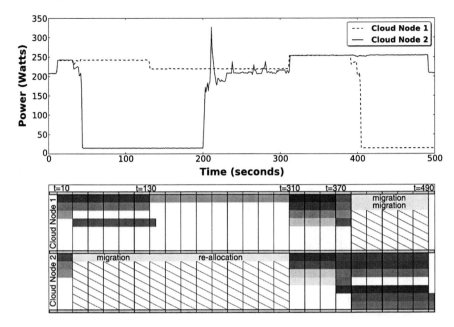

**Fig. 7.11**  Green scenario with round-robin scheduling

During the seventh job, one can notice that a running VM with a *cpuburn* inside consumes about 10 watts, which represents about 5% of the idle consumption of the Cloud node. At time $t = 390$, two new VM migrations are performed, and they draw small peaks on the power consumption curves. From that time, Cloud node 1 is switched off, and Cloud node 2 has six running VMs. We observe that the fifth and the sixth VMs cost nothing (no additional energy consumption compared to the period with only four VMs) as explained in Sect. 7.3.2 as this Cloud node has four cores. This experiment shows that GOC greatly takes advantage of the idle periods by using its green enforcement solutions: VM migration and shut down. Job reallocation also allows us to avoid on/off cycles. Significant amounts of energy are saved.

The green scenario with unbalanced scheduling is presented in Fig. 7.12. For the green scenario, the unbalanced scheduling is more in favor of energy savings. Indeed, two migrations less are needed than with the round-robin scheduling, and all the first burst jobs are allocated to Cloud node 1 allowing Cloud node 2 to stay off. In fact, this is the general case for all the scenarios (Fig. 7.13): the unbalanced scheduling is more energy-efficient since it naturally achieves a better consolidation on fewer nodes.

Summarized results of all the experiments are shown in Fig. 7.13. As expected, the green scenario which illustrates GOC behavior is less energy consuming for the both scheduling scenarios. The balancing scenario is more energy consuming than the basic scheme for unbalanced scheduling. This phenomenon is due to numerous migrations for the balancing scenario that cancel out the benefits of these migrations.

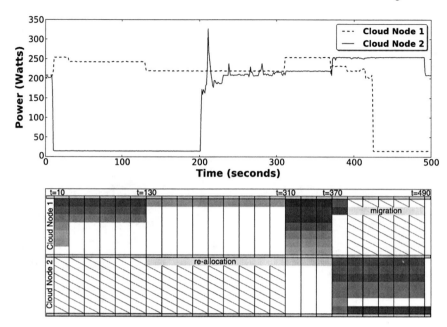

**Fig. 7.12** Green scenario with unbalanced scheduling

**Fig. 7.13** Comparison results

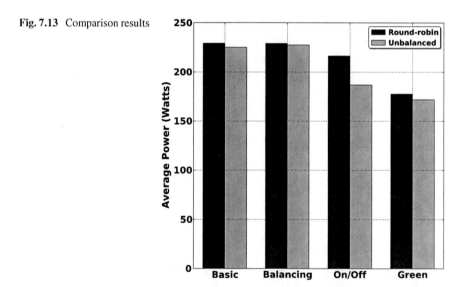

The noticeable figure is that the green scenario with unbalanced scheduling consumes 25% less electricity than the basic scenario which shows the current Clouds administration. This figure obviously depends on the Cloud usage.

However, it shows that great energy savings are achievable with minor impacts on the utilization: small delays occur due to migrations (less than 5 seconds for these experiments) and to physical node boots (2 minutes for our experimental platform nodes). These delays can be improved by using better VM live-migration techniques [13] and by using faster booting techniques (like suspend to disk and suspend to RAM techniques).

## 7.6 Conclusion

As Clouds are more and more broadly used, the Cloud computing ecosystem becomes a key challenge with strong repercussions. It is therefore urgent to analyse and to encompass all the stakeholders related to the Cloud's energy usage in order to design adapted framework solutions. It is also essential to increase the energy-awareness of users in order to reduce their $CO_2$ footprint induced by their utilizations of Cloud infrastructures. This matter leads us to measure and to study the electrical cost of basic operations concerning the VM management in Clouds, such as the boot, the halt, the inactivity and the launching of a sample cpuburn application. These observations show that VMs do not consume energy when they are idle and that booting and halting VMs produce small power peaks but are not really costly in terms of time and energy.

As Clouds are on the way to becoming an essential element of future Internet, new technologies have emerged for increasing their capacity to provide on-demand XaaS (everything as a service) resources in a pay-as-you-use fashion. Among these new technologies, live migration seems to be a really promising approach allowing on-demand resource provisioning and consolidation. We have studied the contribution that the use of this technique could constitute in terms of energy efficiency. VM live migration, although its performance can be improved, is reliable, fast, and requires little energy compared to the amount it can save.

These first measurements have allowed us to propose an original software framework, the Green Open Cloud (GOC) which aims at controlling and optimizing the energy consumed by the overall Cloud infrastructure. This energy management is based on different techniques:

- energy monitoring of the Cloud resources with energy sensor for taking efficient management decisions;
- displaying the energy logs on the Cloud portal to increase the energy-awareness;
- switching off unused resources to save energy;
- utilization of a proxy to carry out network presence of switched-off resources and thus provide inter-operability with any kind of RMS;
- use of VM migration to consolidate the load in fewer resources;
- prediction of the next resource usage to ensure responsiveness;
- giving green advice to users in order to aggregate resources' reservations.

The GOC framework embeds features that, in our opinion, will be part of the next-generation Clouds, such as advance reservations, which enable a more flexible and planned resource management, and VM live-migration, which allows for great workload consolidation.

Further, the GOC framework was the subject of an experimental validation. The validation is made with two different scheduling algorithms to prove GOC's adaptivity to any kind of Resource Management System (RMS). Four scenarios are studied to compare the GOC behavior with basic Cloud's RMS workings. Energy measurements on these scenarios show that GOC can save up to 25% of the energy consumed by a basic Cloud resource management. These energy savings are also reflected in the operational costs of the Cloud infrastructures.

This promising work shows that important energy savings, and thus $CO_2$ footprint reductions are achievable in future Clouds at the cost of minor performance degradations. There is still room for improvement to increase the user's energy-awareness, which is the main leverage to trigger a real involvement from the Cloud providers and designers in order to reduce the electricity demand of Clouds and contribute to a more sustainable ICT industry.

# References

1. Citrix xenserver. URL http://citrix.com/English/ps2/products/product.asp?contentID=683148
2. Make it green—cloud computing and its contribution to climate change. Greenpeace international (2010)
3. Barham, P., Dragovic, B., Fraser, K., Hand, S., Harris, T., Ho, A., Neugebauer, R., Pratt, I., Warfield, A.: Xen and the art of virtualization. In: 19th ACM Symposium on Operating Systems Principles (SOSP '03), pp. 164–177. ACM, New York (2003). doi:10.1145/945445. 945462
4. Box, G.E.P., Jenkins, G.M., Reinsel, G.C.: Time Series Analysis: Forecasting and Control, 3rd edn. Prentice-Hall International, New York (1994)
5. Chase, J.S., Anderson, D.C., Thakar, P.N., Vahdat, A.M., Doyle, R.P.: Managing energy and server resources in hosting centers. In: 18th ACM Symposium on Operating Systems Principles (SOSP '01), pp. 103–116. ACM, Banff (2001)
6. Clark, C., Fraser, K., Hand, S., Hansen, J.G., Jul, E., Limpach, C., Pratt, I., Warfield, A.: Live migration of virtual machines. In: NSDI'05: Proceedings of the 2nd Conference on Symposium on Networked Systems Design & Implementation, Berkeley, CA, USA, pp. 273–286 (2005)
7. Da-Costa, G., Gelas, J.P., Georgiou, Y., Lefèvre, L., Orgerie, A.C., Pierson, J.M., Richard, O., Sharma, K.: The green-net framework: energy efficiency in large scale distributed systems. In: HPPAC 2009: High Performance Power Aware Computing Workshop in Conjunction with IPDPS 2009, Roma, Italy (2009)
8. Doyle, R.P., Chase, J.S., Asad, O.M., Jin, W., Vahdat, A.M.: Model-based resource provisioning in a Web service utility. In: 4th Conference on USENIX Symposium on Internet Technologies and Systems (USITS'03), p. 5. USENIX Association, Berkeley (2003)
9. Fan, X., Weber, W.D., Barroso, L.A.: Power provisioning for a warehouse-sized computer. In: ISCA '07: Proceedings of the 34th Annual International Symposium on Computer Architecture, New York, NY, USA, pp. 13–23 (2007). doi:10.1145/1250662.1250665
10. Fontán, J., Vázquez, T., Gonzalez, L., Montero, R.S., Llorente, I.M.: OpenNEbula: the open source virtual machine manager for cluster computing. In: Open Source Grid and Cluster Software Conference—Book of Abstracts, San Francisco, USA (2008)
11. Hamm, S.: With Sun, IBM Aims for Cloud Computing Heights (26 March 2009). URL http://www.businessweek.com/magazine/content/09_14/b4125034196164.htm?chan= magazine+channel_news

12. Harizopoulos, S., Shah, M.A., Meza, J., Ranganathan, P.: Energy efficiency: the new holy grail of data management systems research. In: Fourth Biennial Conference on Innovative Data Systems Research (CIDR) (2009). URL http://www-db.cs.wisc.edu/cidr/cidr2009/Paper_112.pdf

13. Hirofuchi, T., Nakada, H., Ogawa, H., Itoh, S., Sekiguchi, S.: A live storage migration mechanism over wan and its performance evaluation. In: VTDC '09: Proceedings of the 3rd International Workshop on Virtualization Technologies in Distributed Computing, pp. 67–74. ACM, New York (2009). doi:10.1145/1555336.1555348

14. Hotta, Y., Sato, M., Kimura, H., Matsuoka, S., Boku, T., Takahashi, D.: Profile-based optimization of power performance by using dynamic voltage scaling on a PC cluster. In: IPDPS (2006). doi:10.1109/IPDPS.2006.1639597

15. Iosup, A., Dumitrescu, C., Epema, D., Li, H., Wolters, L.: How are real grids used? the analysis of four grid traces and its implications. In: 7th IEEE/ACM International Conference on Grid Computing (2006)

16. Jung, G., Joshi, K.R., Hiltunen, M.A., Schlichting, R.D., Pu, C.: A cost-sensitive adaptation engine for server consolidation of multitier applications. In: 10th ACM/IFIP/USENIX International Conference on Middleware (Middleware 2009), pp. 1–20. Springer, New York (2009)

17. Kalman, R.E.: A new approach to linear filtering and prediction problems. Trans. ASME, J. Basic Eng. **82**(Series D), 35–45 (1960)

18. Kalyvianaki, E., Charalambous, T., Hand, S.: Self-adaptive and self-configured CPU resource provisioning for virtualized servers using Kalman filters. In: 6th International Conference on Autonomic Computing (ICAC 2009), pp. 117–126. ACM, New York (2009). doi:10.1145/1555228.1555261

19. Kim, M., Noble, B.: Mobile network estimation. In: 7th Annual International Conference on Mobile Computing and Networking (MobiCom 2001), pp. 298–309. ACM, New York (2001). doi:10.1145/381677.381705

20. Kusic, D., Kephart, J.O., Hanson, J.E., Kandasamy, N., Jiang, G.: Power and performance management of virtualized computing environments via lookahead control. In: 5th International Conference on Autonomic Computing (ICAC 2008), pp. 3–12. IEEE Computer Society, Washington (2008). doi:10.1109/ICAC.2008.31

21. Lefèvre, L., Orgerie, A.C.: Designing and evaluating an energy efficient cloud. J. Supercomput. **51**(3), 352–373 (2010)

22. Liu, J., Zhao, F., Liu, X., He, W.: Challenges towards elastic power management in Internet data centers. In: 29th IEEE International Conference on Distributed Computing Systems Workshops (ICDCSW 2009), pp. 65–72. IEEE Computer Society, Washington (2009). doi:10.1109/ICDCSW.2009.44

23. Miyoshi, A., Lefurgy, C., Van Hensbergen, E., Rajamony, R., Rajkumar, R.: Critical power slope: understanding the runtime effects of frequency scaling. In: ICS'02: Proceedings of the 16th International Conference on Supercomputing, pp. 35–44. ACM, New York (2002). doi:10.1145/514191.514200

24. Moore, J., Chase, J., Ranganathan, P., Sharma, R.: Making scheduling "cool": temperature-aware workload placement in data centers. In: USENIX Annual Technical Conference (ATEC 2005), pp. 5–5. USENIX Association, Berkeley (2005)

25. Nathuji, R., Schwan, K.: VirtualPower: Coordinated power management in virtualized enterprise systems. In: 21st ACM SIGOPS Symposium on Operating Systems Principles (SOSP 2007), pp. 265–278. ACM, New York (2007). doi:10.1145/1294261.1294287

26. Nurmi, D., Wolski, R., Crzegorczyk, C., Obertelli, G., Soman, S., Youseff, L., Zagorodnov, D.: Eucalyptus: a technical report on an elastic utility computing architecture linking your programs to useful systems. Technical report 2008-10, Department of Computer Science, University of California, Santa Barbara, California, USA (2008)

27. Orgerie, A.C., Lefèvre, L., Gelas, J.P.: Chasing gaps between bursts: towards energy efficient large scale experimental grids. In: PDCAT 2008: The Ninth International Conference on Parallel and Distributed Computing, Applications and Technologies, pp. 381–389. Dunedin, New Zealand (2008)

28. Orgerie, A.C., Lefèvre, L., Gelas, J.P.: Save watts in your grid: green strategies for energy-aware framework in large scale distributed systems. In: 14th IEEE International Conference on Parallel and Distributed Systems (ICPADS), Melbourne, Australia, pp. 171–178 (2008)

29. Patterson, M., Costello, D., Grimm, P., Loeffler, M.: Data center TCO: a comparison of high-density and low-density spaces. In: Thermal Challenges in Next Generation Electronic Systems (THERMES 2007) (2007). URL http://isdlibrary.intel-dispatch.com/isd/114/datacenterTCO_WP.pdf

30. Sharma, R., Bash, C., Patel, C., Friedrich, R., Chase, J.: Balance of power: dynamic thermal management for internet data centers. IEEE Internet Comput. 9(1), 42–49 (2005). doi:10.1109/MIC.2005.10

31. Silicon Valley Leadership Group: Data center energy forecast. White Paper (2008). URL svlg.org/campaigns/datacenter/docs/DCEFR_report.pdf

32. Singh, T., Vara, P.K.: Smart metering the clouds. In: IEEE International Workshop on Enabling Technologies, pp. 66–71 (2009). doi:10.1109/WETICE.2009.49

33. Snowdon, D.C., Ruocco, S., Heiser, G.: Power management and dynamic voltage scaling: myths and facts. In: Proceedings of the 2005 Workshop on Power Aware Real-time Computing (2005)

34. Srikantaiah, S., Kansal, A., Zhao, F.: Energy aware consolidation for cloud computing. In: Proceedings of HotPower '08 Workshop on Power Aware Computing and Systems (2008). URL http://www.usenix.org/events/hotpower08/tech/full_papers/srikantaiah/srikantaiah/_html/

35. Subramanyam, S., Smith, R., van den Bogaard, P., Zhang, A.: Deploying Web 2.0 applications on Sun servers and the OpenSolaris operating system. Sun BluePrints 820-7729-10, Sun Microsystems (2009)

36. Talaber, R., Brey, T., Lamers, L.: Using Virtualization to Improve Data Center Efficiency. Tech. rep., The Green Grid (2009)

37. Tatezono, M., Maruyama, N., Matsuoka, S.: Making wide-area, multi-site MPI feasible using Xen VM. In: Workshop on Frontiers of High Performance Computing and Networking (held with ISPA 2006). LNCS, vol. 4331, pp. 387–396. Springer, Berlin (2006)

38. Travostino, F., Daspit, P., Gommans, L., Jog, C., de Laat, C., Mambretti, J., Monga, I., van Oudenaarde, B., Raghunath, S., Wang, P.Y.: Seamless live migration of virtual machines over the MAN/WAN. Future Gener. Comput. Syst. 22(8), 901–907 (2006). doi:10.1016/j.future.2006.03.007

39. Urgaonkar, B., Shenoy, P., Chandra, A., Goyal, P., Wood, T.: Agile dynamic provisioning of multi-tier internet applications. ACM Trans. Auton. Adapt. Syst. 3(1), 1–39 (2008). doi:10.1145/1342171.1342172

40. Verma, A., Ahuja, P., Neogi, A.: pMapper: power and migration cost aware application placement in virtualized systems. In: ACM/IFIP/USENIX 9th International Middleware Conference (Middleware 2008), pp. 243–264. Springer, Berlin (2008). doi:10.1007/978-3-540-89856-6_13

# Chapter 8
# Jungle Computing: Distributed Supercomputing Beyond Clusters, Grids, and Clouds

Frank J. Seinstra, Jason Maassen, Rob V. van Nieuwpoort, Niels Drost,
Timo van Kessel, Ben van Werkhoven, Jacopo Urbani, Ceriel Jacobs,
Thilo Kielmann, and Henri E. Bal

**Abstract** In recent years, the application of high-performance and distributed computing in scientific practice has become increasingly wide spread. Among the most widely available platforms to scientists are clusters, grids, and cloud systems. Such infrastructures currently are undergoing revolutionary change due to the integration of many-core technologies, providing orders-of-magnitude speed improvements for selected compute kernels. With high-performance and distributed computing sys-

F.J. Seinstra (✉) · J. Maassen · R.V. van Nieuwpoort · N. Drost · T. van Kessel ·
B. van Werkhoven · J. Urbani · C. Jacobs · T. Kielmann · H.E. Bal
Department of Computer Science, Vrije Universiteit, De Boelelaan 1081A, 1081 HV Amsterdam,
The Netherlands
e-mail: fjseins@cs.vu.nl

J. Maassen
e-mail: jason@cs.vu.nl

R.V. van Nieuwpoort
e-mail: rob@cs.vu.nl

N. Drost
e-mail: niels@cs.vu.nl

T. van Kessel
e-mail: timo@cs.vu.nl

B. van Werkhoven
e-mail: ben@cs.vu.nl

J. Urbani
e-mail: jacopo@cs.vu.nl

C. Jacobs
e-mail: ceriel@cs.vu.nl

T. Kielmann
e-mail: kielmann@cs.vu.nl

H.E. Bal
e-mail: bal@cs.vu.nl

M. Cafaro, G. Aloisio (eds.), *Grids, Clouds and Virtualization,*
Computer Communications and Networks,
DOI 10.1007/978-0-85729-049-6_8, © Springer-Verlag London Limited 2011

tems thus becoming more heterogeneous and hierarchical, programming complexity is vastly increased. Further complexities arise because urgent desire for scalability and issues including data distribution, software heterogeneity, and ad hoc hardware availability commonly force scientists into *simultaneous use of multiple platforms* (e.g., clusters, grids, and clouds used concurrently). A true *computing jungle*.

In this chapter we explore the possibilities of enabling efficient and transparent use of *Jungle Computing Systems* in everyday scientific practice. To this end, we discuss the fundamental methodologies required for defining programming models that are tailored to the specific needs of scientific researchers. Importantly, we claim that many of these fundamental methodologies already exist today, as integrated in our Ibis high-performance distributed programming system. We also make a case for the urgent need for easy and efficient Jungle Computing in scientific practice, by exploring a set of state-of-the-art application domains. For one of these domains, we present results obtained with Ibis on a real-world Jungle Computing System. The chapter concludes by exploring fundamental research questions to be investigated in the years to come.

## 8.1 Introduction

It is widely recognized that Information and Communication Technologies (ICTs) have revolutionized the everyday practice of science [6, 54]. Whereas in earlier times scientists spent a lifetime recording and analyzing observations by hand, in many research laboratories today much of this work has been automated. The benefits of automation are obvious: it allows researchers to increase productivity by increasing efficiency, to improve quality by reducing error, and to cope with increasing scale—enabling scientific treatment of topics that were previously impossible to address.

As a direct result of automation, in many research domains the rate of scientific progress is now faster than ever before [11, 14, 59]. Importantly, however, the rate of progress itself puts further demands on the automation process. The availability of ever larger amounts of observational data, for example, directly leads to increasing needs for computing, networking, and storage. As a result, for many years, the scientific community has been one of the major driving forces behind state-of-the-art developments in supercomputing technologies (e.g., see [58]).

Although this self-stimulating process indeed allows scientists today to study more complex problems than ever before, it has put a severe additional burden on the scientists themselves. Many scientists have to rely on arguably the most complex computing architectures of all—i.e., high-performance and distributed computing systems in their myriad of forms. To effectively exploit the available processing power, a thorough understanding of the complexity of such systems is essential. As a consequence, the number of scientists capable of using such systems effectively (if at all) is relatively low [44].

Despite the fact that there is an obvious need for programming solutions that allow scientists to obtain high-performance and distributed computing both efficiently and transparently, real solutions are still lacking [5, 24]. Worse even, the

high-performance and distributed computing landscape is currently undergoing revolutionary change. Traditional clusters, grids, and cloud systems are more and
more equipped with state-of-the-art many-core technologies (e.g., Graphics Processing Units or GPUs [31, 32]). Although these devices often provide orders-of-
magnitude speed improvements, they make computing platforms more heterogeneous and hierarchical—and vastly more complex to program and use.

Further complexities arise in everyday practice. Given the ever increasing need
for compute power, and due to additional issues including data distribution, software
heterogeneity, and ad hoc hardware availability, scientists are commonly forced to
apply multiple clusters, grids, clouds, and other systems *concurrently*—even for
single applications. In this chapter we refer to such a simultaneous combination of
heterogeneous, hierarchical, and distributed computing resources as a *Jungle Computing System*.

In this chapter we explore the possibilities of enabling efficient and transparent
use of Jungle Computing Systems in everyday scientific practice. To this end, we
focus on the following research question:

> What are the fundamental methodologies required for defining programming models that
> are tailored to the specific needs of scientific researchers and that match state-of-the-art
> developments in high-performance and distributed computing architectures?

We will claim that many of these fundamental methodologies already exist and have
been integrated in our Ibis software system for high-performance and distributed
computing [4]. In other words: Jungle Computing is not just a visionary concept; to
a large extent, we already adhere to its requirements today.

This chapter is organized as follows. In Sect. 8.2 we discuss several architectural revolutions that are currently taking place—leading to the new notion of Jungle Computing. Based on these groundbreaking developments, Sect. 8.3 defines the
general requirements underlying transparent programming models for Jungle Computing Systems. Section 8.4 discusses the Ibis programming system and explores
to what extent Ibis adheres to the requirements of Jungle Computing. Section 8.5
sketches a number of emerging problems in various science domains. For each domain, we will stress the need for Jungle Computing solutions that provide transparent speed and scalability. For one of these domains, Sect. 8.6 evaluates the Ibis
platform on a real-world Jungle Computing System. Section 8.7 introduces a number of fundamental research questions to be investigated in the coming years and
concludes.

## 8.2 Jungle Computing Systems

When grid computing was introduced over a decade ago, its foremost visionary
aim (or "promise") was to provide *efficient and transparent (i.e., easy-to-use) wall-
socket computing over a distributed set of resources* [18]. Since then, many other
distributed computing paradigms have been introduced, including peer-to-peer computing [25], volunteer computing [57], and, more recently, cloud computing [15].

These paradigms all share many of the goals of grid computing, eventually aiming to provide end-users with access to distributed resources (ultimately even at a worldwide scale) with as little effort as possible.

These new distributed computing paradigms have led to a diverse collection of resources available to research scientists, including stand-alone machines, cluster systems, grids, clouds, desktop grids, etc. Extreme cases in terms of computational power further include mobile devices at the low end of the spectrum and supercomputers at the top end.

If we take a step back and look at such systems from a high-level perspective, then all of these systems share important common characteristics. Essentially, *all* of these systems consist of a number of basic compute nodes, each having local memories and each capable of communicating over a local or wide-area connection. The most prominent differences are in the semantic and administrative organization, with many systems providing their own middlewares, programming interfaces, access policies, and protection mechanisms [4].

Apart from the increasing diversity in the distributed computing landscape, the "basic compute nodes" mentioned above currently are undergoing revolutionary change as well. General-purpose CPUs today have multiple compute cores per chip, with an expected increase in the years to come [40]. Moreover, special-purpose chips (e.g., GPUs [31, 32]) are now combined or even integrated with CPUs to increase performance by orders-of-magnitude (e.g., see [28]).

The many-core revolution is already affecting the field of high-performance and distributed computing today. One interesting example is the Distributed ASCI Supercomputer 4 (DAS-4), which is currently being installed in The Netherlands. This successor to the earlier DAS-3 system (see www.cs.vu.nl/das3/) will consist of six clusters located at five different universities and research institutes, with each cluster being connected by a dedicated and fully optical wide-area connection. Notably, each cluster will also contain a variety of many-core "add-ons" (including a.o. GPUs and FPGAs), making DAS-4 a highly diverse and heterogeneous system. The shear number of similar developments currently taking place the world over indicates that many-cores are rapidly becoming an irrefutable additional component of high-performance and distributed systems.

With clusters, grids, and clouds thus being equipped with multi-core processors and many-core "add-ons," systems available to scientists are becoming increasingly hard to program and use. Despite the fact that the programming and efficient use of many-cores is known to be hard [31, 32], this is not the only—or most severe—problem. With the increasing heterogeneity of the underlying hardware, the efficient mapping of computational problems onto the "bare metal" has become vastly more complex. Now more than ever, programmers must be aware of the potential for parallelism *at all levels of granularity*.

But the problem is even more severe. Given the ever increasing desire for speed and scalability in many scientific research domains, the use of a *single* high-performance computing platform is often not sufficient. The need to access multiple platforms concurrently from within a single application often is due to the impossibility of reserving a sufficient number of compute nodes at once in a single multiuser

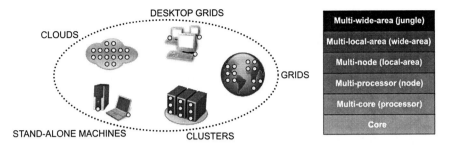

**Fig. 8.1** *Left*: A "worst-case" Jungle Computing System as perceived by scientific end-users, simultaneously comprising any number of clusters, grids, clouds, and other computing platforms. *Right*: Hierarchical view of a Jungle Computing System

system. Moreover, additional issues such as the distributed nature of the input data, the heterogeneity of the software pipeline being applied, and the ad hoc availability of the required computing resources, further indicate a need for computing across multiple, and potentially very diverse, platforms. For all of these reasons, in this chapter we make the following claim:

> Many research scientists (now and in the near future) are being forced to apply multiple clusters, grids, clouds, and other systems *concurrently*—even for executing single applications.

We refer to such a simultaneous combination of heterogeneous, hierarchical, and distributed computing resources as a *Jungle Computing System* (see Fig. 8.1).

The abovementioned claim is not new. As part of the European Grid Initiative (EGI [55]), for example, it has been stressed that the integrated application of clusters, grids, and clouds in scientific computing is a key component of the research agenda for the coming years [9]. Similarly, Microsoft Research has advocated the integrated use of grids and clouds [16]. Further European research efforts in this direction are taking place in COST Action IC0805 (ComplexHPC: Open European Network for High-Performance Computing in Complex Environments [56]).

Compared to these related visionary ideas, the notion of a Jungle Computing System is more all-encompassing. It exposes *all* computing problems that scientists today can be (and often are) confronted with. Even though we do not expect most (or even any) research scientists to have to deal with the "worst-case" scenario depicted in Fig. 8.1, we do claim that—in principle—*any* possible subset of this figure represents a realistic scenario. Hence, if we can define the fundamental methodologies required to solve the problems encountered in the worst-case scenario, we ensure that our solution applies to all possible scenarios.

## 8.3  Jungle Computing: Requirements and Methodologies

Although Jungle Computing Systems and grids are not identical (i.e., the latter being constituent components of the former), a generic answer to our overall research question introduced in Sect. 8.1 is given by the "founding fathers of the grid." In [18], Foster et al. indicate that one of the main aims of grid computing is to deliver *transparent* and potentially *efficient* computing, even at a worldwide scale. This aim extends to Jungle Computing as well.

It is well known that adhering to the general requirements of transparency and efficiency is a hard problem. Although rewarding approaches exist for specific application types (i.e., work-flow driven problems [29, 47] and parameter sweeps [1]), solutions for more general applications types (e.g., involving irregular communication patterns) do not exist today. This is unfortunate, as advances in optical networking allow for a much larger class of distributed (Jungle Computing) applications to run efficiently [53].

We ascribe this rather limited use of grids and other distributed systems—or the lack of efficient and transparent programming models—to the intrinsic complexities of distributed (Jungle) computing systems. Programmers often are required to use low-level programming interfaces that change frequently. Also, they must deal with system- and software heterogeneity, connectivity problems, and resource failures. Furthermore, managing a running application is hard, because the execution environment may change dynamically as resources come and go. All these problems limit the acceptance of the many distributed computing technologies available today.

In our research we aim to overcome these problems and to drastically simplify the programming and deployment of distributed supercomputing applications—without limiting the set of target hardware platforms. Importantly, our philosophy is that Jungle Computing applications should be developed on a local workstation and simply be launched from there. This philosophy directly leads to a number of fundamental requirements underlying the notion of Jungle Computing. In the following we will give a high-level overview of these requirements and indicate how these requirements are met with in our Ibis distributed programming system.

### 8.3.1 Requirements

The abovementioned general requirements of transparency and efficiency are unequal quantities. The requirement of transparency decides whether an end-user is capable of using a Jungle Computing System *at all*, while the requirement of efficiency decides whether the use is sufficiently *satisfactory*. In the following we will therefore focus mainly on the transparency requirements. We will simply assume that, once the requirement of transparency is fulfilled, efficiency is a derived property that can be obtained a.o. by introducing "intelligent" optimization techniques, application domain-specific knowledge, etc.

In our view, for full transparency, the end-user must be shielded from *all* issues that complicate the programming and use of Jungle Computing Systems in comparison with the programming and use of a desktop computer. To this end, methodologies must be available that provide transparent support for:

- *Resource independence.* In large-scale Jungle Computing Systems heterogeneity is omnipresent, to the effect that applications designed for one system are generally guaranteed to fail on others. This problem must be removed by hiding the physical characteristics of resources from end-users.
- *Middleware independence and interoperability.* As many resources already have at least one middleware installed, Jungle Computing applications must be able to use (or: interface with) such *local middlewares*. To avoid end-users having to implement a different interface for each local middleware (and to enhance portability), it is essential to have available a single high-level interface on top of *all* common middleware systems. As multiple distributed resources may use different middlewares, some form of interoperability between these middlewares must be ensured as well.
- *Robust connectivity and globally unique resource naming.* Getting distributed applications to execute at all in a Jungle Computing System is difficult. This is because firewalls, transparent renaming of IP addresses, and multihoming (machines with multiple addresses) can severely complicate or limit the ability of resources to communicate. Moreover, in many cases no direct connection with certain machines is allowed at all. Despite solutions that have been provided for firewall issues (e.g., NetIbis [10], Remus [46]), integrated solutions must be made available that remove connectivity problems altogether. At the same time, and in contrast to popular communication libraries such as MPI, each resource must be given a globally unique identifier.
- *Malleability.* In a Jungle Computing System, the set of available resources may change, e.g., because of reservations ending. Jungle Computing software must support malleability, correctly handling resources joining and leaving.
- *System-level fault-tolerance.* Given the many independent parts of a large-scale Jungle Computing System, the chance of resource failures is high. Jungle Computing software must be able to handle such failures in a graceful manner. Failures should not hinder the functioning of the entire system, and failing resources should be detected and, if needed (and possible), replaced.
- *Application-level fault-tolerance.* The capacity of detecting resource failures, and replacing failed resources, is essential functionality for any realistic Jungle Computing System. However, this functionality in itself cannot guarantee the correct continuation of running applications. Hence, restoring the state of applications that had been running on a failed resource is a further essential requirement. Such functionality is generally to be implemented either in the application itself or in the runtime system of the programming model with which an application is implemented. Support for application-level fault-tolerance in the lower levels of the software stack can be limited to failure detection and reporting.
- *Parallelization.* For execution on any Jungle Computing system, it is generally up to the programmer to identify the available parallelism in a problem at hand. For the programmer—generally a domain expert with limited or no expertise in distributed supercomputing—this is often an insurmountable problem. Clearly, programming models must be made available that hide most (if not all) of the inherent complexities of parallelization.

- *Integration with external software.* It is unrealistic to assume that a single all-encompassing software system would adhere to all needs of all projected users. In many scientific research domains there is a desire for integrating "black box" legacy codes, while the expertise or resources to rewrite such codes into a newly required format or language are lacking. Similarly, it is essential to be able to integrate system-level software (e.g., specialized communication libraries) and architecture-specific compute kernels (e.g., CUDA-implemented algorithms for GPU-execution). While such "linking up" with existing and external software partially undermines our "write-and-go" philosophy, this property is essential for a software system for Jungle Computing to be of any use to general scientific researchers.

Our list of requirements is by no means complete; it merely consists of a *minimal* set of methodologies that—in combination—fulfill our high-level requirement of transparency. Further requirements certainly also exist, including support for co-allocation, security, large-scale distributed data management, noncentralized control, runtime adaptivity, the handling of quality-of-service (QoS) constraints, and runtime monitoring and visualization of application behavior. These are secondary requirements, however, and are not discussed further in this chapter.

## 8.4 Ibis

The Ibis platform (see also www.cs.vu.nl/ibis/) aims to combine all of the stated fundamental methodologies into a single integrated programming system that applies to *any* Jungle Computing System (see Fig. 8.2). Our open-source software system provides high-level, architecture- and middleware-independent interfaces that allow for (transparent) implementation of efficient applications that are robust to faults and dynamic variations in the availability of resources. To this end, Ibis consists of a rich software stack that provides all functionality that is traditionally associated with *programming languages* and *communication libraries* on the one hand and *operating systems* on the other. More specifically, Ibis offers an integrated, layered solution, consisting of two subsystems: the High-Performance Programming System and the Distributed Deployment System.

### 8.4.1 The Ibis High-Performance Programming System

The Ibis High-Performance Programming System consists of (1) the IPL, (2) the programming models, and (3) SmartSockets, described below.

(1) *The Ibis Portability Layer (IPL):* The IPL is at the heart of the Ibis High-Performance Programming System. It is a communication library which is written entirely in Java, so it runs on any platform that provides a suitable Java Virtual Machine (JVM). The library is typically shipped with an application (as Java jar

**Fig. 8.2** Integration of the required methodologies in the Ibis software system. See also http://www.cs.vu.nl/ibis/

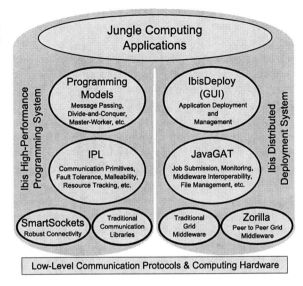

files) such that no preinstalled libraries need to be present at any destination machine. The IPL provides a range of communication primitives (partially comparable to those provided by libraries such as MPI), including point-to-point and multicast communication, and streaming. It applies efficient protocols that avoid copying and other overheads as much as possible, and uses *bytecode rewriting* optimizations for efficient transmission.

To deal with real-world Jungle Computing Systems, in which resources can crash and can be added or deleted, the IPL incorporates a globally unique resource naming scheme and a runtime mechanism that keeps track of the available resources. The mechanism, called Join-Elect-Leave (JEL [13]), is based on the concept of signaling, i.e., notifying the application when resources have Joined or Left the computation. JEL also includes Elections, to select resources with a special role. The IPL contains a centralized implementation of JEL, which is sufficient for static (closed-world) programming models (like MPI), and a more scalable distributed implementation based on gossiping.

The IPL has been implemented on top of the socket interface provided by the JVM and on top of our own SmartSockets library (see below). Irrespective of the implementation, the IPL can be used "out of the box" on any system that provides a suitable JVM. In addition, the IPL can exploit specialized native libraries, such as a Myrinet device driver (MX) if it exists on the target system. Further implementations of the IPL exist, on top of MPI, and on the Android smart phone platform.

(2) *Ibis Programming Models:* The IPL can be (and has been) used directly to write applications, but Ibis also provides several higher-level programming models on top of the IPL, including (1) an implementation of the MPJ standard, i.e., an MPI version in Java, (2) Satin, a divide-and-conquer model, described below, (3) Remote Method Invocation (RMI), an object-oriented form of Remote Procedure Call, (4) Group Method Invocation (GMI), a generalization of RMI to group com-

munication, (5) Maestro, a fault-tolerant and self-optimizing data-flow model, and (6) Jorus, a user transparent parallel programming model for multimedia applications discussed in Sect. 8.6.

Arguably, the most transparent model of these is Satin [60], a divide-and-conquer system that automatically provides fault-tolerance and malleability. Satin recursively splits a program into subtasks and then waits until the subtasks have been completed. At runtime a Satin application can adapt the number of nodes to the degree of parallelism, migrate a computation away from overloaded resources, remove resources with slow communication links, and add new resources to replace resources that have crashed. As such, Satin is one of the few systems that provides transparent programming capabilities in dynamic systems.

(3) *SmartSockets:* To run a parallel application on multiple distributed resources, it is necessary to establish network connections. In practice, however, a variety of connectivity problems exists that make communication difficult or even impossible, such as firewalls, Network Address Translation (NAT), and multihoming. It is generally up to the application user to solve such connectivity problems manually.

The SmartSockets library aims to solve connectivity problems automatically, with little or no help from the user. SmartSockets integrates existing and novel solutions, including reverse connection setup, STUN, TCP splicing, and SSH tunneling. SmartSockets creates an overlay network by using a set of interconnected support processes, called *hubs*. Typically, hubs are run on the front-end of a cluster. Using gossiping techniques, the hubs automatically discover to which other hubs they can establish a connection. The power of this approach was demonstrated in a worldwide experiment: in 30 realistic scenarios SmartSockets always was capable of establishing a connection, while traditional sockets only worked in six of these [30].

Figure 8.3 shows an example using three cluster systems. Cluster A is open and allows all connections. Cluster B uses a firewall that only allows outgoing connections. In cluster C only the front-end machine is reachable. No direct communication is possible between the nodes and the outside world. By starting a hub on each of the front-end machines and providing the location of the hub on cluster A to each of them, the hubs will automatically connect as shown in Fig. 8.3. The arrows between

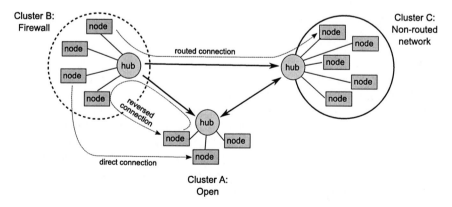

**Fig. 8.3** Example connection setup with SmartSockets

the hubs depict the direction in which the connection can be established. Once a connection is made, it can be used in both directions.

The nodes of the clusters can now use the hubs as an overlay network when no direct communication is possible. For example, when a node of cluster B tries to connect to a node of cluster A, a direct connection can immediately be created. A connection from A to B, however, cannot be established directly. In this case, SmartSockets will use the overlay network to exchange control messages between the nodes to reverse the direction of connection setup. As a result, the desired (direct) connection between the nodes of A and B can still be created. The nodes of cluster C are completely unreachable from the other clusters. In this case SmartSockets will create a *virtual connection*, which routes messages over the overlay network.

The basic use of SmartSockets requires manual initialization of the network of hubs. This task, however, is performed automatically and transparently by IbisDeploy, our top-level deployment system, described in the next section.

## 8.4.2 The Ibis Distributed Deployment System

The Ibis Distributed Application Deployment System consists of a software stack for deploying and monitoring applications, once they have been written. The software stack consists of (1) the JavaGAT, (2) IbisDeploy, and (3) Zorilla.

(1) *The Java Grid Application Toolkit (JavaGAT):* Today, distributed system programmers generally have to implement their applications against a grid middleware API that changes frequently, is low-level, unstable, and incomplete [33]. The JavaGAT solves these problems in an integrated manner. JavaGAT offers high-level primitives for developing and running applications, *independent* of the middleware that implements this functionality [51]. The primitives include access to remote data, start remote jobs, support for monitoring, steering, user authentication, resource management, and storing of application-specific data. The JavaGAT uses an extensible architecture, where *adaptors* (plugins) provide access to the different middlewares.

The JavaGAT integrates multiple middleware systems with different and incomplete functionality into a single, consistent system, using a technique called *intelligent dispatching*. This technique dynamically forwards application calls on the JavaGAT API to one or more middleware adaptors that implement the requested functionality. The selection process is done at runtime and uses policies and heuristics to automatically select the best available middleware, enhancing portability. If an operation fails, the intelligent dispatching feature will automatically select and dispatch the API call to an alternative middleware. This process continues until a middleware successfully performs the requested operation. Although this flexibility comes at the cost of some runtime overhead, compared to the cost of the operations themselves, this is often negligible. For instance, a Globus job submission takes several seconds, while the overhead introduced by the JavaGAT is less than 10 milliseconds. The additional semantics of the high-level API, however, can introduce some overhead. If a file is copied, for example, the JavaGAT first checks if the destination already exists or is a directory. These extra checks may be costly because they require remote operations. Irrespective of these overheads, JavaGAT is essential to support our "write-and-go" philosophy: it allows programmers to ignore low-level systems peculiarities and to focus on solving domain-specific problems instead.

The JavaGAT does not provide a new user/key management infrastructure. Instead, its security interface provides generic functionality to store and manage security information such as usernames and passwords. Also, the JavaGAT provides a mechanism to restrict the availability of security information to certain middleware systems or remote machines. Currently, JavaGAT supports many different middleware systems, such as Globus, Unicore, gLite, PBS, SGE, KOALA, SSH, GridSAM, EC2, ProActive, GridFTP, HTTP, SMB/CIFS, and Zorilla.

(2) *IbisDeploy:* Even though JavaGAT is a major step forward to simplifying application deployment, its API still requires the programmer to think in terms of middleware operations. Therefore, the Ibis Distributed Deployment System provides a simplified and generic API, implemented on top of the JavaGAT, and an additional GUI, the IbisDeploy system.

The IbisDeploy API is a thin layer on top of the JavaGAT API that initializes JavaGAT in the most commonly used ways and that lifts combinations of multiple JavaGAT calls to a higher abstraction level. For example, if one wants to run a distributed application written in Ibis, a network of SmartSockets hubs must be started manually. IbisDeploy takes over this task in a fully transparent manner. Also, to run (part of) an Ibis application on a remote machine, one of the necessary steps is to manually upload the actual program code and related libraries to that machine. IbisDeploy transparently deals with such prestaging (and poststaging) actions as well.

The IbisDeploy GUI (see Fig. 8.4) allows a user to manually load resources and applications at any time. As such, multiple Jungle Computing applications can be started using the same graphical interface. The IbisDeploy GUI also allows the user to add new resources to a running application (by providing contact information such as host address and user credentials), and to pause and resume applications. All runtime settings can be saved and reused in later experiments.

(3) *Zorilla:* Most existing middleware APIs lack coscheduling capabilities and do not support fault tolerance and malleability. To overcome these problems, Ibis provides Zorilla, a lightweight peer-to-peer middleware that runs on any Jungle Computing System. In contrast to traditional middleware, Zorilla has no central components and is easy to set up and maintain. Zorilla supports fault tolerance and malleability by implementing all functionality using peer-to-peer techniques. If resources used by an application are removed or fail, Zorilla is capable of automatically finding replacement resources. Zorilla is specifically designed to easily combine resources in multiple administrative domains.

A Zorilla system is made up of a collection of nodes running on all resources, connected by a peer-to-peer network (see Fig. 8.5). Each node in the system is completely independent and implements all functionality required for a middleware, including the handling of the submission of jobs, running jobs, storing of files, etc. Each Zorilla node has a number of local resources. This may simply be the machine

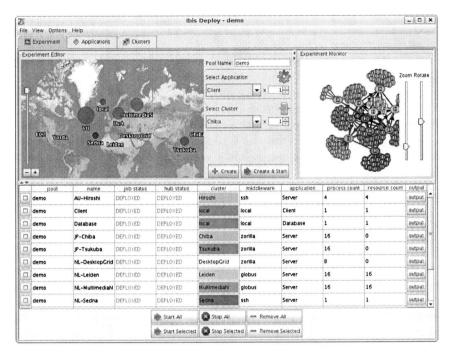

**Fig. 8.4** The IbisDeploy GUI that enables runtime loading of applications and resources (*top middle*) and keeping track of running processes (*bottom half*). *Top left* shows a world map of the locations of available resources; *top right* shows the SmartSockets network of hubs. See also http://www.cs.vu.nl/ibis/demos.html

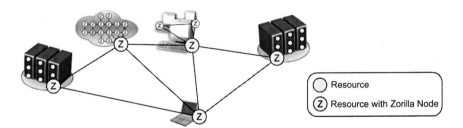

**Fig. 8.5** Example Zorilla peer-to-peer resource-pool on top of a Jungle Computing System consisting of two clusters, a desktop grid, a laptop, and cloud resources (e.g., acquired via Amazon EC2). On the clusters, a Zorilla node is run on the headnode, and Zorilla interacts with the local resources via the local scheduler. On the desktop grid and the cloud, a Zorilla node is running on each resource, since no local middleware capable of scheduling jobs is present on these systems

it is running on, consisting of one or more processor cores, memory, and data storage. Alternatively, a node may provide access to other resources, for instance, to all machines in a cluster. Using the peer-to-peer network, all Zorilla nodes tie together into one big distributed system. Collectively, nodes implement the required global functionality such as resource discovery, scheduling, and distributed data storage, all using peer-to-peer techniques.

To create a resource pool, a Zorilla daemon process must be started on each participating machine. Also, each machine must be given the address of at least one other machine, to set up a connection. Jobs can be submitted to Zorilla using the JavaGAT or, alternatively, using a command line interface. Zorilla then allocates the requested number of resources and schedules the application, taking user-defined requirements (like memory size) into account. The combination of virtualization and peer-to-peer techniques thus makes it very easy to deploy applications with Zorilla.

### 8.4.3 Ibis User Community

Ibis has been used to run a variety of real-life applications like multimedia computing (see Sect. 8.6), spectroscopic data processing (by the Dutch Institute for Atomic and Molecular Physics), human brain scan analysis (with the Vrije Universiteit Medical Center), automatic grammar learning, and many others. Also, Ibis has been applied successfully in an implementation of the industry-strength SAT4J SAT-solver. In addition, Ibis has been used by external institutes to build high-level programming systems, such as a workflow engine for astronomy applications in D-grid (Max-Planck-Institute for Astrophysics) and a grid file system (University of Erlangen–Nürnberg), or to enhance existing systems, such as KOALA (Delft University of Technology), ProActive (INRIA), Jylab (University of Patras), and Grid Superscalar (Barcelona Supercomputer Center). Moreover, Ibis has won prizes in international competitions, such as the International Scalable Computing Challenge at CCGrid 2008 and 2010 (for scalability), the International Data Analysis Challenge for Finding Supernovae at IEEE Cluster/Grid 2008 (for speed and fault-tolerance), and the Billion Triples Challenge at the 2008 International Semantic Web Conference (for general innovation).

### 8.4.4 Ibis versus the Requirements of Jungle Computing

From the general overview of the Ibis platform it should be clear that our software system adheres to most (if not all) of the requirements of Jungle Computing introduced in Sect. 8.3. Resource independence is obtained by relying on JVM virtualization, while the JavaGAT provides us with middleware independence and interoperability. Robust connectivity and globally unique resource naming is taken care of by the SmartSockets library and the Ibis Portability Layer (IPL), respectively. The need for malleability and system-level fault-tolerance is supported by the resource tracking mechanisms of the Join-Elect-Leave (JEL) model, which is an integral part of the Ibis Portability Layer. Application-level fault-tolerance can be built on top of the system-level fault-tolerance mechanisms provided by the IPL. For a number of programming models, in particular Satin, we have indeed done so in a fully transparent manner. The need for transparent parallelization is also fulfilled through a number of programming models implemented on top of the IPL. Using the Satin model, parallelization is obtained automatically for divide-and-conquer applications. Similarly, using the Jorus model, data parallel multimedia computing applications can be implemented in a fully user transparent manner. Finally, integration with legacy codes and system-level software is achieved through JNI (Java Native Interface) link ups, in the case of system level software through so-called adaptor interfaces (plugins). We will further highlight the linking up with many-core specific compute kernels implemented using CUDA in Sect. 8.6.

## 8.5 The Need for Jungle Computing in Scientific Practice

As stated in Sect. 8.1, the scientific community has automated many daily activities, in particular to *speed up* the generation of results and to *scale up* to problem sizes that better match the research questions at hand. Whether it be in the initial process of data collection or in later stages including data filtering, analysis, and storage, the desire for speed and scalability can occur in any phase of the scientific process.

In this section we describe a number of urgent and realistic problems occurring in four representative science domains: Multimedia Content Analysis, Semantic Web, Neuroinformatics, and Remote Sensing. Each description focuses on the societal significance of the domain, the fundamental research questions, and the unavoidable need for applying Jungle Computing Systems.

### 8.5.1 Multimedia Content Analysis

Multimedia Content Analysis (MMCA) considers all aspects of the automated extraction of knowledge from multimedia data sets [7, 45]. MMCA applications (both real-time and offline) are rapidly gaining importance along with recent deployment of publicly accessible digital TV archives and surveillance cameras in public locations. In a few years, MMCA will be a problem of phenomenal proportions, as digital video may produce high data rates, and multimedia archives steadily run into Petabytes of storage. For example, the analysis of the TRECVID data set [42], consisting of 184 hours of video, was estimated to take over 10 years on a fast sequential computer. While enormous in itself, this is insignificant compared to the 700,000 hours of TV data archived by the Dutch Institute for Sound and Vision. Moreover, distributed sets of surveillance cameras generate even larger quantities of data.

Clearly, for emerging MMCA problems, there is an urgent need for speed and scalability. Importantly, there is overwhelming evidence that large-scale distributed supercomputing indeed can push forward the state-of-the-art in MMCA. For example, in [42] we have shown that a distributed system involving hundreds of massively communicating resources covering the entire globe indeed can bring efficient

**Fig. 8.6** Real-time (*left*) and offline (*right*) distributed multimedia computing. The real-time application constitutes a visual object recognition task, in which video frames obtained from a robot camera are processed on a set of compute clusters. Based on the calculated scene descriptions, a database of learned objects is searched. Upon recognition the robot reacts accordingly. For this system, we obtained a "most visionary research award" at AAAI 2007. A video presentation is available at http://www.cs.vu.nl/~fjseins/aibo.html. The offline application constitutes our TRECVID system, in which low-level semantic concepts (e.g., "sky", "road", "greenery") are extracted and combined into high-level semantic concepts (e.g., "cars driving on a highway"), using the same set of compute clusters. For this system, we obtained a "best technical demo award" at ACM Multimedia 2005

solutions for real-time and offline problems (see Fig. 8.6). Notably, our Ibis-based solution to the real-time problem of Fig. 8.6 has won First Prize in the International Scalable Computing Challenge at CCGrid in 2008.

### 8.5.2 Semantic Web

The Semantic Web [23, 49] is a groundbreaking development of the World Wide Web in which the semantics of information is defined. Through these semantics machines can "understand" the Web, allowing querying and reasoning over Web information gathered from different sources. The Semantic Web is based on specifications that provide formal descriptions of concepts, terms, and relationships within a given knowledge domain. Examples include annotated medical datasets containing. e.g., gene sequence information (linkedlifedata.com) and structured information derived from Wikipedia (dbpedia.org).

Even though the Semantic Web domain is still in its infancy, it already faces problems of staggering proportions [17]. Today, the field is dealing with huge distributed repositories containing billions of facts and relations (see also Fig. 8.7)— with an expected exponential growth in the years to come. As current Semantic Web reasoning systems do not scale to the requirements of emerging applications, it has been acknowledged that there is an urgent need for a distributed platform for massive reasoning that will remove the speed and scalability barriers [17]. Notably, our preliminary work in this direction has resulted in prize-winning contributions to the Billion Triples Challenge at the International Semantic Web Conference in 2008 [2] and the International Scalable Computing Challenge at CCGrid in 2010 [48].

**Fig. 8.7** Visualization of relationships between concepts and terms extracted through automatic reasoning. As the full dataset is far too large, this example visualization focuses on one term only. This is a feat in itself, as the filtering out of such data from the full dataset is a complex search problem. Notably, our prize-winning reasoner is 60 times faster and deals with 10 times more data than any existing approach

### 8.5.3 Neuroinformatics

Neuroinformatics encompasses the analysis of experimental neuroscience data for improving existing theories of brain function and nervous system growth. It is well known that activity dynamics in neuronal networks depend largely on the pattern of synaptic connections between neurons [12]. It is not well understood, however, how neuronal morphology and the local processes of neurite outgrowth and synapse formation influence global connectivity. To investigate these issues, there is a need for simulations that can generate large neuronal networks with realistic neuronal morphologies.

Due to the high computational complexity, existing frameworks (such as NET-MORPH [27]) can simulate networks of up to a few hundred neurons only (Fig. 8.8, left). With the human brain having an estimated 100 billion neurons, a vast scaling up of the simulations is urgent. While the neuroinformatics domain is not new to the use of supercomputer systems (e.g., the Blue Brain Project [21]), speed and scale requirements—as well as algorithmic properties—dictate distributed supercomputing at an unprecedented scale. Our preliminary investigation of the NETMORPH system has shown significant similarities with a specific subset of $N$-body problems that require adaptive runtime employment of compute resources. We have ample

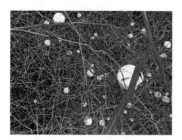  

**Fig. 8.8** *Left*: NETMORPH-generated network, with cell bodies, axons, dendrites and synaptic connections embedded in 3D space. *Right*: AVIRIS hyperspectral image data, with location of fires in World Trade Center

experience with such problems [41, 60], which require many of the advanced capabilities of the Ibis system.

## 8.5.4 Remote Sensing

Remotely sensed hyperspectral imaging is a technique that generates hundreds of images corresponding to different wavelengths channels for the same area on the surface of the Earth [20]. For example, NASA is continuously gathering image data with satellites such as the Jet Propulsion Laboratory's Airborne Visible-Infrared Imaging Spectrometer (AVIRIS [22]). The resulting hyperspectral data cubes consist of high-dimensional pixel vectors representing spectral signatures that uniquely characterize the underlying objects [8]. Processing of these cubes is highly desired in many application domains, including environmental modeling and risk/hazard prediction (Fig. 8.8, right).

With emerging instruments generating in the order of 1 Gbit/s (e.g., ESA's FLEX [39]), data sets applied in real applications easily approach the petascale range. Given the huge computational demands, parallel and distributed solutions are essential, at all levels of granularity. As a result, in the last decade the use of compute clusters for applications in remote sensing has become commonplace [38]. These approaches are proven beneficial for a diversity of problems, including target detection and classification [35], and automatic spectral unmixing and endmember extraction [37]. Depending on the complexity and dimensionality of the analyzed scene, however, the computational demands of many remote sensing methodologies still limit their use in time-critical applications. For this reason, there is an emerging trend in mapping remote sensing functionality onto multi- and many-core hardware and in combining the resulting compute kernels with existing solutions for clusters and distributed systems [36].

### 8.5.5  A Generalized View

The science domains described above are not unique in their needs for speed and scalability. These domains are simply the ones for which we have gained experience over the years. Certainly, and as clearly shown in [3], the list of domains that we could have included here is virtually endless.

Importantly, however, the set of described domains covers a wide range of *application types*, with some being time-constrained and compute-intensive, and others being offline and data-intensive. Also, for all of these domains, Jungle Computing solutions already exist today (at least to a certain extent)—with each using a variety of distributed computing systems (even at a worldwide scale), and some including the use of many-core hardware. As shown in Sect. 8.6, some of these solutions are implemented purely in Ibis, while others constitute a "mixed-language" solution with legacy codes and specialized compute kernels being integrated with Ibis software. As stated, many of these Ibis-implemented solutions have won prizes and awards at international venues, each for a different reason: speed, scalability, fault-tolerance, and general innovation. With our ultimate goal of developing transparent and efficient tools for scientific domain experts in mind, it is relevant to note that several of these results have been obtained with little or no help from the Ibis team.

This brings us to the reasons for the *unavoidable need* for using Jungle Computing Systems in these and other domains, as claimed in this chapter. While these reasons are manyfold, we will only state the most fundamental ones here. First, in many cases research scientists need to acquire a number of compute nodes that cannot be provided by a single system alone—either because the problem at hand is too big, or because other applications are being run on part of the available hardware. In these cases, concurrently acquiring nodes on multiple systems often is the only route to success. In other cases, calculations need to be performed on distributed data sets that cannot be moved—either because of size limitations, or due to reasons of privacy, copyright, security, etc. In these cases, it is essential to (transparently) move the calculations to where the data is, instead of vice versa. Furthermore, because many scientists have to rely on legacy codes or compute kernels for special-purpose hardware, parts of the processing pipeline may only run on a limited set of available machines. In case the number of such specialized codes becomes large, acquiring resources from many different resources simply is unavoidable.

It is important to realize that the use of a variety of distributed resources (certainly at a worldwide scale) is *not* an active desire or end-goal of any domain expert. For all of the above (and other) reasons, research scientists today are simply being *forced* into using Jungle Computing Systems. It is up to the field of high-performance and distributed computing to understand the fundamental research questions underlying this problem, to develop the fundamental methodologies solving each of these research questions, and to combine these methodologies into efficient and transparent programming models and tools for end-users.

## 8.6 Jungle Computing Experiments

In this section we describe a number of experiments that illustrate the functionality and performance of the Ibis system. We focus on the domain of Multimedia Content Analysis, introduced in Sect. 8.5.1. Our discussion starts with a description of an Ibis-implemented programming model, called Jorus, which is specifically targeted toward researchers in the MMCA domain [42].

For the bulk of the experiments, we use the Distributed ASCI Supercomputer 3 (DAS-3, www.cs.vu.nl/das3), a five cluster/272-dual node distributed system located at four universities in The Netherlands. The clusters are largely homogeneous, but there are differences in the number of cores per machine, the clock frequency, and the internal network. In addition, we use a small GPU-cluster, called Lisa, located at SARA (Stichting Academisch Rekencentrum Amsterdam). Although the traditional part of the Lisa cluster is much larger, the system currently has a total of six Quad-core Intel Xeon 2.50 GHz nodes available, each of which is equipped with two Nvidia Tesla M1060 graphics adaptors with 240 cores and 4 GBytes of device memory. Next to DAS-3 and Lisa, we use additional clusters in Chicago (USA), Chiba and Tsukuba (InTrigger, Japan), and Sydney (Australia), an Amazon EC2 Cloud system (USA, East Region), as well as a desktop grid and a single stand-alone machine (both Amsterdam, The Netherlands). Together, this set of machines constitutes a real-world Jungle Computing System as defined earlier. Most of the experiments described below are supported by a video presentation, which is available at http://www.cs.vu.nl/ibis/demos.html.

### 8.6.1 High-Performance Distributed Multimedia Analysis with Jorus

The Jorus programming system is an Ibis-implemented *user transparent* parallelization tool for the MMCA domain. Jorus is the next generation implementation of our library-based Parallel-Horus system [42], which was implemented in C++ and MPI. Jorus and Parallel-Horus allow programmers to implement *data parallel* multimedia applications as fully sequential programs. Apart from the obvious differences between the Java and C++ languages, the Jorus and Parallel-Horus APIs are identical to that of a popular *sequential* programming system, called Horus [26]. Similar to other frameworks [34], Horus recognizes that a small set of *algorithmic patterns* can be identified that covers the bulk of all commonly applied multimedia computing functionality.

Jorus and Parallel-Horus include patterns for functionality such as unary and binary pixel operations, global reduction, neighborhood operation, generalized convolution, and geometric transformations (e.g., rotation, scaling). Recent developments include patterns for operations on large datasets and patterns on increasingly important derived data structures, such as feature vectors. For reasons of efficiency, all Jorus and Parallel-Horus operations are capable of adapting to the performance characteristics of the cluster computer at hand, i.e., by being flexible in the partitioning of data structures. Moreover, it was realized that it is not sufficient to consider parallelization of library operations *in isolation*. Therefore, the programming systems incorporate a runtime approach for communication minimization (called *lazy parallelization*) that automatically parallelizes a fully sequential program at runtime by inserting communication primitives and additional memory management operations whenever necessary [43].

Earlier results obtained with Parallel-Horus for realistic multimedia applications have shown the feasibility of the applied approach, with data parallel performance consistently being found to be optimal with respect to the abstraction level of message passing programs [42]. Notably, Parallel-Horus was applied in earlier NIST TRECVID benchmark evaluations for content-based video retrieval and played a crucial role in achieving top-ranking results in a field of strong international competitors [42, 45]. Moreover, and as shown in our evaluation below, extensions to Jorus and Parallel-Horus that allow for services-based distributed multimedia computing have been applied successfully in large-scale distributed systems, involving hundreds of massively communicating compute resources covering the entire globe [42]. Finally, while the current Jorus implementation realizes data parallel execution on cluster systems in a fully user transparent manner, we are also working on a cluster implementation that results in combined data and task parallel execution [50].

## 8.6.2 Experiment 1: Fine-Grained Parallel Computing

To start our evaluation of Ibis for Jungle Computing, we first focus on Ibis communication on a single traditional cluster system. To be of any significance for Jungle Computing applications, it is essential for Ibis' communication performance to compare well to that of the MPI message passing library—the de facto standard for high-performance cluster computing applications. Therefore, in this first experiment we will focus on fine-grained data-parallel image and video analysis, implemented using our Jorus programming model. For comparison, we also report results obtained for Parallel-Horus (in C++/MPI) [42].

In particular, we investigate the data-parallel analysis of a single video frame in a typical MMCA application (as also shown in the left half of Fig. 8.6). The application implements an advanced object recognition algorithm developed in the Dutch MultimediaN project [42]. At runtime, so-called *feature vectors* are extracted from the video data, each describing local properties like color and shape. The analysis of a single video frame is a data-parallel task executed on a cluster. When using multiple clusters, data-parallel calculations over consecutive frames are executed concurrently in a task-parallel manner.

The Jorus implementation intentionally mimics the original Parallel-Horus version as close as possible. Hence, Jorus implements several collective communication operations, such as scatter, gather, and all-reduce. Other application specific communication steps, such as the exchange of nonlocal image data between nodes (known as *BorderExchange*) are implemented in a manner resembling point-to-point communication. More importantly, the Ibis runtime environment has been set up for *closed-world* execution, meaning that the number of compute nodes is fixed for the duration of the application run. This approach voids all of Ibis' fault-tolerance and malleability capabilities, but it shows that *Ibis can be used easily to mimic any MPI-style application*.

By default, Ibis uses TCP for communication, but the user can indicate which communication protocol to use. In our experiments the MX (Myrinet Express) protocol is available for the DAS-3 Myri-10G high-speed interconnects. In the following we will therefore report results for four different versions of the multimedia application using a single physical network: two versions in Java/Ibis (one communicating over MX and one over TCP) and two versions in C++/MPI (again, one for each protocol). Of these four versions, the C++/MPI/MX version is expected to be the fastest, as C++ is perceived to be generally faster than Java and MX is designed specifically for communication over Myri-10G.

Figure 8.9(a) presents the results for the DAS-3 cluster at the Vrije Universiteit, obtained for a video frame of size $1024 \times 768$ pixels. The sequential Java version is about 20% slower than the sequential C++ version. This performance drop is well within acceptable limits for a "compile-once, run everywhere" application executing inside a virtual machine. Also, MX does not significantly outperform TCP. Clearly, the communication patterns applied in our application do not specifically favor MX. More importantly, the two Ibis versions are equally fast and show similar speedup characteristics compared to their MPI-based counterparts.

Further evidence of the feasibility of Ibis for parallel computing is shown in Fig. 8.9(b). The graph on the left shows the normalized speed of three versions of the application compared to the fastest C++/MPI/MX version. It shows that the relative drop in performance is rather stable at 20 to 25%, which is attributed to JVM overhead. The graph on the right presents the cost of communication relative to the overall execution time. Clearly, the relative parallelization overheads of Ibis and MPI are almost identical. These are important results, given the increased flexibility and much wider applicability of the Ibis system.

### 8.6.3 Experiment 2: User Transparent MMCA on GPU-Clusters

Essentially, Jorus extends the original sequential Horus library by introducing a thin layer right in the heart of the small set of algorithmic patterns, as shown in Fig. 8.10. This layer uses the IPL to communicate image data and other structures among the different nodes in a cluster. In the most common case, a digital image is *scattered* throughout the parallel system, so that each compute node is left with a *partial image*. Apart from the need for additional pre- and post-communication steps (such as the common case of *border handling* in convolution operations), the *sequential compute kernels* as also available in the original Horus system are now applied to each partial image.

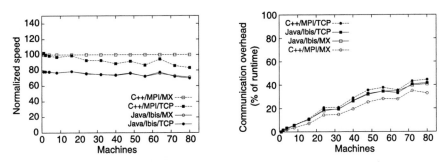

| # Nodes | C++/MPI/MX | Java/Ibis/MX | C++/MPI/TCP | Java/Ibis/TCP |
|---|---|---|---|---|
| 1 | 7.935 | 10.132 | 7.751 | 10.132 |
| 2 | 3.959 | 5.051 | 3.982 | 5.050 |
| 4 | 1.989 | 2.543 | 2.019 | 2.553 |
| 8 | 1.012 | 1.315 | 1.045 | 1.313 |
| 16 | 0.540 | 0.687 | 0.546 | 0.686 |
| 24 | 0.377 | 0.496 | 0.407 | 0.500 |
| 32 | 0.289 | 0.387 | 0.311 | 0.385 |
| 40 | 0.258 | 0.347 | 0.291 | 0.350 |
| 48 | 0.233 | 0.305 | 0.255 | 0.307 |
| 56 | 0.204 | 0.282 | 0.235 | 0.281 |
| 64 | 0.188 | 0.247 | 0.199 | 0.242 |
| 72 | 0.179 | 0.245 | 0.209 | 0.249 |
| 80 | 0.168 | 0.235 | 0.201 | 0.240 |

(a) Performance and speedup characteristics of all application versions.

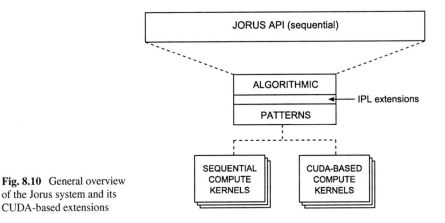

(b) Normalized speed and communication overhead of all application versions.

**Fig. 8.9**   Results obtained on DAS-3 cluster at Vrije Universiteit, Amsterdam

From a software engineering perspective, the fact that the IPL extensions "touch" the sequential implementation of the algorithmic patterns in such a minimal way provides Jorus with the important properties of sustainability and easy extensibility. In the process of extending Jorus for GPU-based execution, we have obtained a similar minimal level of intrusiveness: we left the thin communication layer as it is and introduced CUDA-based alternatives to the sequential compute kernels that imple-

**Fig. 8.10**   General overview of the Jorus system and its CUDA-based extensions

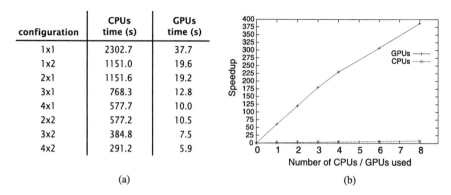

| configuration | CPUs time (s) | GPUs time (s) |
|---|---|---|
| 1x1 | 2302.7 | 37.7 |
| 1x2 | 1151.0 | 19.6 |
| 2x1 | 1151.6 | 19.2 |
| 3x1 | 768.3 | 12.8 |
| 4x1 | 577.7 | 10.0 |
| 2x2 | 577.2 | 10.5 |
| 3x2 | 384.8 | 7.5 |
| 4x2 | 291.2 | 5.9 |

(a)                                                                 (b)

**Fig. 8.11** (a) Runtimes in seconds for different configurations in the number of CPUs and GPUs; (b) User-perceived speedup compared to single CPU execution. Results obtained on Lisa cluster, SARA, Amsterdam

ment the algorithmic patterns (see bottom half of Fig. 8.10). In this manner, Jorus and CUDA are able to work in concert, allowing the use of multiple GPUs on the same node and on multiple nodes simultaneously, simply by creating one Jorus process for each GPU. In other words, with this approach, we obtain a system that can execute sequential Jorus applications in data parallel fashion, while exploiting the power of GPU hardware. The details of the CUDA-implemented compute kernels are beyond the scope of this chapter and are discussed further in [52].

Our second experiment comprises the computationally demanding problem of *line detection* in images and video streams [19]. Although not identical to our object recognition problem discussed above, the set of algorithmic patterns applied in the Jorus implementation is quite similar. We have executed this GPU-enabled application on the Lisa cluster. In our experiments we use many different configurations, each of which is denoted differently by the number of nodes and CPUs/GPUs used. For example, measurements involving one compute node and one Jorus process are denoted by $1 \times 1$. Likewise, $4 \times 2$ means that 4 nodes are used with 2 Jorus processes executing on each node. For the CUDA-based executions, the latter case implies the concurrent use of 8 GPUs.

Figure 8.11(a) shows the total runtimes for several configurations using either CPUs or GPUs for the execution of the original and CUDA-implemented compute kernels. The presented execution times include the inherently sequential part of the application, which consists mainly of reading and writing the input and output images. For CPU-only execution, the execution time reduces linearly; using 8 CPUs gives a speedup of 7.9. These results are entirely in line with earlier speedup characteristics reported in [44] for much larger cluster systems.

The GPU-extensions to the Jorus system show a dramatic performance improvement in comparison with the original version. Even in the $1 \times 1$ case, the total execution time is reduced by a factor of 61.2. The speedup gained from executing on a GPU cluster, compared to a traditional cluster, clearly demonstrates why traditional clusters are now being extended with GPUs as accelerators. As shown in

Fig. 8.11(b), our application executing on 4 nodes with 2 Jorus processes per node, experiences a speedup of 387 with GPU-extensions. Notably, when using the Jorus programming model, these speedup results are obtained without requiring *any* parallelization effort from the application programmer.

## 8.6.4 Experiment 3: Jungle Computing at a World-Wide Scale

After having discussed some the capabilities and performance characteristics of the Ibis system for traditional cluster systems and emerging GPU-clusters, we will now turn our attention to the use of Ibis for worldwide execution on a large variety of computing resources. For this purpose, we reconsider our object recognition problem of Sect. 8.6.2.

As stated, when using multiple distributed resources, with Jorus it is possible to concurrently perform multiple data-parallel calculations over consecutive video frames in a task-parallel manner. This is achieved by wrapping the data parallel analysis in a *multimedia server* implementation. At runtime, client applications can then upload an image or video frame to such a server and receive back a recognition result. In case multiple servers are available, a client can use these simultaneously for subsequent image frames, in effect resulting in task-parallel employment of data-parallel services.

As shown in our demonstration video (see www.cs.vu.nl/ibis/demos.html), we use IbisDeploy to start a client and an associated database of learned objects on a local machine and to deploy four data-parallel multimedia servers, each on a different DAS-3 cluster (using 64 machines in total). All code is implemented in Java and Ibis, and compiled on the local machine. No application codes are initially installed on any other machine.

Our distributed application shows the simultaneous use of multiple Ibis environments. Whilst the data-parallel execution runs in a closed-world setting, the distributed extensions are set up for open-world execution to allow resources to be added and removed at runtime. For this application, the additional use of resources indeed is beneficial: where the use of a single multimedia server results in a client-side processing rate of approximately 1.2 frames per second, the simultaneous use of 2 and 4 clusters leads to linear speedups at the client side with 2.5 and 5 frames/s, respectively.

We continue the experiment by employing several additional clusters, an Amazon EC2 cloud, a local desktop grid, and a local standalone machine. With this worldwide set of machines, we now use a variety of middlewares simultaneously (i.e., Globus, Zorilla, and SSH) from within a single application. Although not included in the demonstration, at a lower video resolution the maximum obtained frame-rate is limited only by the speed of the camera, meaning that (soft) real-time multimedia computing at a worldwide scale has become a reality.

When running at this worldwide scale, a variety of connectivity problems is automatically circumvented by SmartSockets. As almost all of the applied resources

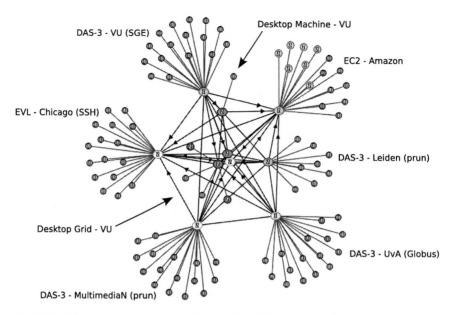

**Fig. 8.12** Visualization of resources used in a worldwide Jungle Computing experiment, showing all nodes and the SmartSockets overlay network between these. Nodes marked $Z$ represent Zorilla nodes running on either a frontend machine or on a resource itself. Nodes marked $I$ represent instances of a running (Ibis) application

have more than one IP address, SmartSockets automatically selects the appropriate address. Also, SmartSockets automatically creates SSH tunnels to connect with the clusters in Japan and Australia. These systems would be unreachable otherwise. Finally, the SmartSockets network of hubs avoids further problems due to firewalls and NATs. In case we would remove the hub network, we only would have access to those machines that allow direct connections *to* and *from* our client application. As our client is behind a firewall, two-way connectivity is possible only within the same administrative domain, which includes the local desktop grid, the standalone machine, and one of the DAS-3 clusters. Clearly, the absence of SmartSockets would by-and-large reduce our worldwide distributed system to a small set of local machines. A visualization of a SmartSockets overlay network is shown in Fig. 8.12.

To illustrate Ibis' fault-tolerance mechanisms, the video also shows an experiment where an entire multimedia server crashes. The Ibis resource tracking system notices the crash, and signals this event to other parts of the application. After some time, the client is notified, to the effect that the crashed server is removed from the list of available servers. The application continues to run, as the client keeps on forwarding video frames to all remaining servers.

The demonstration also shows the use of the multimedia servers from a smartphone running the Android operating system. For this platform, we have written a separate Ibis-based client, which can upload pictures taken with the phone's camera and receive back a recognition result. Running the full application on the smartphone itself is not possible due to CPU and memory limitations. Based on a stripped down, yet very inaccurate, version that *does* run on the smartphone, we estimate that recognition for $1024 \times 768$ images would take well over 20 minutes. In contrast, when the smartphone uploads a picture to one of the multimedia servers, it obtains a result in about 3 seconds. This result clearly shows the potential of Ibis to open up the field of mobile computing for compute-intensive applications. Using IbisDeploy, it is even possible to deploy the entire distributed application as described from the smartphone itself.

## 8.7  Conclusions and Future Work

In this chapter we have argued that, while the need for speed and scalability in everyday scientific practice is omnipresent and increasing, the resources employed by end-users are often more diverse than those contained in a single cluster, grid, or cloud system. In many realistic scientific research areas, domain experts are being forced into concurrent use of multiple clusters, grids, clouds, desktop grids, stand-alone machines, and more. Writing applications for such *Jungle Computing Systems* has become increasingly difficult, in particular with the integration of many-core hardware technologies.

The aim of the Ibis platform is to drastically simplify the programming and deployment of Jungle Computing applications. To achieve this, Ibis integrates solutions to many of the fundamental problems of Jungle Computing in a single modular programming and deployment system, written entirely in Java. Ibis has been used for many real-world Jungle Computing applications and has won awards and prizes in very diverse international competitions.

Despite the successes, and the fact that—to our knowledge—Ibis is the only integrated system that offers an efficient and transparent solution for Jungle Computing, further progress is urgent for Ibis to become a viable programming system for everyday scientific practice. One of the foremost questions to be dealt with is whether it is possible to define a set of fundamental building blocks that can describe *any* Jungle Computing application. Such building blocks can be used to express both generic programming models (e.g., pipelining, divide-and-conquer, MapReduce, SPMD), and domain-specific models (e.g., the Jorus model described earlier). A further question is whether all of these models indeed can yield efficient execution on various Jungle Computing Systems. In relation to this is the question whether it is possible to define generic computational patterns that can be reused to express a variety of domain-specific programming models. The availability of such generic patterns would significantly enhance the development of new programming models for unexplored scientific domains.

As we have shown in Sect. 8.6.3, multiple kernels with identical functionality but targeted at different platforms (referred to as *equi-kernels*) often are available. Such kernels are all useful, e.g., due to different scalability characteristics or ad hoc hardware availability. Therefore, an important question is how to transparently integrate (multiple) domain-specific kernels with Jungle Computing programming models and applications. Moreover, how do we transparently decide to schedule a specific kernel when multiple equi-kernels can be executed on various resources? Initially, it would be sufficient to apply random or heuristic-based approaches. For improved performance, however, further solutions must be investigated, including those that take into account the potential benefits of coalescing multiple subsequent kernels, and scheduling these as a single kernel.

Mapping kernels to resources is a dynamic problem. This is because resources may be added or removed, and the computational requirements of kernels may fluctuate over time. Moreover, the mapping may have to take into account optimization under multiple, possibly conflicting, objectives (e.g., speed, productivity, financial costs, energy use). Hence, a further research question is to what extent it is possible to provide runtime support for transparent and dynamic remapping and migration of compute kernels in Jungle Computing Systems. Also, what basic metrics do we need for making self-optimization decisions, and how can we apply these using existing theories of multiobjective optimization?

Despite the need for solutions to all of these (and other) fundamental research questions, the Ibis system was shown to adhere to most (if not all) of the necessary requirements of Jungle Computing introduced in Sect. 8.3. As a result, we conclude that Jungle Computing is not merely a visionary concept; with Ibis, we already deploy Jungle Computing applications on a worldwide scale virtually every day. Anyone can do the same: the Ibis platform is fully open source and can be downloaded for free from http://www.cs.vu.nl/ibis.

# References

1. Abramson, D., Sosic, R., Giddy, J., Hall, B.: Nimrod: a tool for performing parameterised simulations using distributed workstations. In: Proceedings of the 4th IEEE International Symposium on High Performance Distributed Computing (HPDC'95), Pentagon City, USA, pp. 112–121 (1995)
2. Anadiotis, G., Kotoulas, S., Oren, E., Siebes, R., van Harmelen, F., Drost, N., Kemp, R., Maassen, J., Seinstra, F., Bal, H.: MaRVIN: a distributed platform for massive RDF inference. In: Semantic Web Challenge 2008, Held in Conjunction with the 7th International Semantic Web Conference (ISWC 2008), Karlsruhe, Germany (2008)
3. Asanovic, K., Bodik, R., Demmel, J., Keaveny, T., Keutzer, K., Kubiatowicz, J., Morgan, N., Patterson, D., Sen, K., Wawrzynek, J., Wessel, D., Yelick, K.: A view of the parallel computing landscape. Commun. ACM 52(10), 56–67 (2009)
4. Bal, H., Maassen, J., van Nieuwpoort, R., Drost, N., Kemp, R., van Kessel, T., Palmer, N., Wrzesińska, G., Kielmann, T., van Reeuwijk, K., Seinstra, F., Jacobs, C., Verstoep, K.: Real-world distributed computing with ibis. IEEE Comput. 48(8), 54–62 (2010)
5. Butler, D.: The petaflop challenge. Nature 448, 6–7 (2007)
6. Carley, K.: Organizational change and the digital economy: a computational organization science perspective. In: Brynjolfsson, E., Kahin, B. (eds.) Understanding the Digital Economy: Data, Tools, Research, pp. 325–351. MIT Press, Cambridge (2000)
7. Carneiro, G., Chan, A., Moreno, P., Vasconcelos, N.: Supervised learning of semantic classes for image annotation and retrieval. IEEE Trans. Pattern Anal. Mach. Intell. 29(3), 394–410 (2007)
8. Chang, C.I.: Hyperspectral Data Exploitation: Theory and Applications. Wiley, New York (2007)

9. Kranzlmüller, D.: Towards a sustainable federated grid infrastructure for science. In: Keynote Talk, Sixth High-Performance Grid Computing Workshop (HPGC'08), Rome, Italy (2009)
10. Denis, A., Aumage, O., Hofman, R., Verstoep, K., Kielmann, T., Bal, H.: Wide-area communication for grids: an integrated solution to connectivity, performance and security problems. In: Proceedings of the 13th International Symposium on High Performance Distributed Computing (HPDC'04), Honolulu, HI, USA, pp. 97–106 (2004)
11. Dijkstra, E.: On the Phenomenon of Scientific Disciplines (1986). Unpublished Manuscript EWD988; E.W. Dijkstra Archive
12. Douglas, R., Martin, K.: Neuronal circuits in the neocortex. Annu. Rev. Neurosci. **27**, 419–451 (2004)
13. Drost, N., van Nieuwpoort, R., Maassen, J., Seinstra, F., Bal, H.: JEL: unified resource tracking for parallel and distributed applications. Concurr. Comput. Pract. Exp. (2010). doi:10.1002/cpe.1592
14. Editorial: The importance of technological advances. Nature Cell Biology **2**, E37 (2000)
15. Editorial: Cloud computing: clash of the clouds. The Economist (2009)
16. Gagliardi, F.: Grid and cloud computing: opportunities and challenges for e-science. In: Keynote Speech, International Symposium on Grid Computing 2008 (ISCG 2008), Taipei, Taiwan (2008)
17. Fensel, D., van Harmelen, F., Andersson, B., Brennan, P., Cunningham, H., Valle, E.D., Fischer, F., Zhisheng, H., Kiryakov, A., Lee, T.I., Schooler, L., Tresp, V., Wesner, S., Witbrock, M., Ning, Z.: Towards LarKC: a platform for web-scale reasoning. In: Proceedings of the Second International Conference on Semantic Computing (ICSC 2008), Santa Clara, CA, USA, pp. 524–529 (2008)
18. Foster, I., Kesselman, C., Tuecke, S.: The anatomy of the grid: enabling scalable virtual organizations. Int. J. High Perform. Comput. Appl. **15**(3), 200–222 (2001)
19. Geusebroek, J., Smeulders, A., Geerts, H.: A minimum cost approach for segmenting networks of lines. Int. J. Comput. Vis. **43**(2), 99–111 (2001)
20. Goetz, A., Vane, G., Solomon, J., Rock, B.: Imaging spectrometry for earth remote sensing. Science **228**, 1147–1153 (1985)
21. Graham-Rowe, D.: Mission to Build a Simulated Brain Begins. New Scientist (2005)
22. Green, R., Eastwood, M., Sarture, C., Chrien, T., Aronsson, M., Chippendale, B., Faust, J., Pavri, B., Chovit, C., Solis, M., Olah, M.: Imaging spectroscopy and the airborne visible/infrared imaging spectrometer (AVIRIS). Remote Sens. Environ. **65**(3), 227–248 (1998)
23. Hendler, J., Shadbolt, N., Hall, W., Berners-Lee, T., Weitzner, D.: Web science: an interdisciplinary approach to understanding the web. Commun. ACM **51**(7), 60–69 (2008)
24. Hey, T.: The social grid. In: Keynote Talk, OGF20 2007, Manchester, UK (2007)
25. Khan, J., Wierzbicki, A.: Guest editor's introduction; foundation of peer-to-peer computing. Comput. Commun. **31**(2), 187–189 (2008)
26. Koelma, D., Poll, E., Seinstra, F.: Horus C++ reference. Tech. rep., University of Amsterdam, The Netherlands (2002)
27. Koene, R., Tijms, B., van Hees, P., Postma, F., de Ridder, A., Ramakers, G., van Pelt, J., van Ooyen, A.: NETMORPH: a framework for the stochastic generation of large scale neuronal networks with realistic neuron morphologies. Neuroinformatics **7**(3), 195–210 (2009)
28. Lu, P., Oki, H., Frey, C., Chamitoff, G., Chiao, L., Fincke C.M. Foale, E.M. Jr., Tani, D., Whitson, P., Williams, J., Meyer, W., Sicker, R., Au, B., Christiansen, M., Schofield, A., Weitz, D.: Order-of-magnitude performance increases in gpu-accelerated correlation of images from the international space station. J. Real-Time Image Process. (2009)
29. Ludäscher, B., Altintas, I., Berkley, C., Higgins, D., Jaeger, E., Jones, M., Lee, E., Tao, J., Zhao, Y.: Scientific workflow management and the Kepler system. Concurr. Comput. Pract. Exp. **18**(10), 1039–1065 (2005)
30. Maassen, J., Bal, H.: SmartSockets: solving the connectivity problems in grid computing. In: Proceedings of the 16th International Symposium on High Performance Distributed Computing (HPDC'07), Monterey, USA, pp. 1–10 (2007)
31. Manual: Advanced Micro Devices Corporation (AMD). AMD Stream Computing User Guide, Revision 1.1 (2008)

32. Manual: NVIDIA CUDA Complete Unified Device Architecture Programming Guide, v2.0 (2008)

33. Medeiros, R., Cirne, W., Brasileiro, F., Sauvé, J.: Faults in grids: why are they so bad and what can be done about it? In: Proceedings of the 4th International Workshop on Grid Computing, Phoenix, AZ, USA, pp. 18–24 (2003)

34. Morrow, P., Crookes, D., Brown, J., McAleese, G., Roantree, D., Spence, I.: Efficient implementation of a portable parallel programming model for image processing. Concurr. Comput. Pract. Exp. **11**, 671–685 (1999)

35. Paz, A., Plaza, A., Plaza, J.: Comparative analysis of different implementations of a parallel algorithm for automatic target detection and classification of hyperspectral images. In: Proceedings of SPIE Optics and Photonics—Satellite Data Compression, Communication, and Processing V, San Diego, CA, USA (2009)

36. Plaza, A.: Recent developments and future directions in parallel processing of remotely sensed hyperspectral images. In: Proceedings of the 6th International Symposium on Image and Signal Processing and Analysis, Salzburg, Austria, pp. 626–631 (2009)

37. Plaza, A., Plaza, J., Paz, A.: Parallel heterogeneous CBIR system for efficient hyperspectral image retrieval using spectral mixture analysis. Concurr. Comput. Pract. Exp. **22**(9), 1138–1159 (2010)

38. Plaza, A., Valencia, D., Plaza, J., Martinez, P.: Commodity cluster-based parallel processing of hyperspectral imagery. J. Parallel Distrib. Comput. **66**(3), 345–358 (2006)

39. Rasher, U., Gioli, B., Miglietta, F.: FLEX—fluorescence explorer: a remote sensing approach to quantify spatio-temporal variations of photosynthetic efficiency from space. In: Allen, J., et al. (eds.) Photosynthesis. Energy from the Sun: 14th International Congress on Photosynthesis, pp. 1387–1390. Springer, Berlin (2008)

40. Reilly, M.: When multicore isn't enough: trends and the future for multi-multicore systems. In: Proceedings of the Twelfth Annual Workshop on High-Performance Embedded Computing (HPEC 2008), Lexington, MA, USA (2008)

41. Seinstra, F., Bal, H., Spoelder, H.: Parallel simulation of ion recombination in nonpolar liquids. Future Gener. Comput. Syst. **13**(4–5), 261–268 (1998)

42. Seinstra, F., Geusebroek, J., Koelma, D., Snoek, C., Worring, M., Smeulders, A.: High-performance distributed video content analysis with parallel-horus. IEEE Trans. Multimed. **14**(4), 64–75 (2007)

43. Seinstra, F., Koelma, D., Bagdanov, A.: Finite state machine-based optimization of data parallel regular domain problems applied in low-level image processing. IEEE Trans. Parallel Distrib. Syst. **15**(10), 865–877 (2004)

44. Seinstra, F., Koelma, D., Geusebroek, J.: A software architecture for user transparent parallel image processing. Parallel Comput. **28**(7–8), 967–993 (2002)

45. Snoek, C., Worring, M., Geusebroek, J., Koelma, D., Seinstra, F., Smeulders, A.: The semantic pathfinder: using an authoring metaphor for generic multimedia indexing. IEEE Trans. Pattern Anal. Mach. Intell. **28**(10), 1678–1689 (2006)

46. Tan, J., Abramson, D., Enticott, C.: Bridging organizational network boundaries on the grid. In: Proceedings of the 6th IEEE International Workshop on Grid Computing, Seattle, WA, USA, pp. 327–332 (2005)

47. Taylor, I., Wang, I., Shields, M., Majithia, S.: Distributed computing with Triana on the grid. Concurr. Comput. Pract. Exp. **17**(9), 1197–1214 (2005)

48. Urbani, J., Kotoulas, S., Maassen, J., Drost, N., Seinstra, F., van Harmelen, F., Bal, H.: WebPIE: a web-scale parallel inference engine. In: Third IEEE International Scalable Computing Challenge (SCALE2010), Held in Conjunction with the 10th IEEE/ACM International Symposium on Cluster, Cloud and Grid Computing (CCGrid 2010), Melbourne, Australia (2010)

49. van Harmelen, F.: Semantic web technologies as the foundation of the information infrastructure. In: van Oosterom, P., Zlatanove, S. (eds.) Creating Spatial Information Infrastructures: Towards the Spatial Semantic Web. CRC Press, London (2008)

50. van Kessel, T., Drost, N., Seinstra, F.: User transparent task parallel multimedia content analysis. In: Proceedings of the 16th International Euro-Par Conference (Euro-Par 2010), Ischia–Naples, Italy (2010)
51. van Nieuwpoort, R., Kielmann, T., Bal, H.: User-friendly and reliable grid computing based on imperfect middleware. In: Proceedings of the ACM/IEEE International Conference on Supercomputing (SC'07), Reno, NV, USA (2007)
52. van Werkhoven, B., Maassen, J., Seinstra, F.: Towards user transparent parallel multimedia computing on GPU-clusters. In: Proceedings of the 37th ACM IEEE International Symposium on Computer Architecture (ISCA 2010), First Workshop on Applications for Multi and Many Core Processors (A4MMC 2010), Saint Malo, France (2010)
53. Verstoep, K., Maassen, J., Bal, H., Romein, J.: Experiences with fine-grained distributed supercomputing on a 10G testbed. In: Proceedings of the 8th IEEE International Symposium on Cluster Computing and the Grid (CCGrid'08), Lyon, France, pp. 376–383 (2008)
54. Waltz, D., Buchanan, B.: Automating science. Science **324**, 43–44 (2009)
55. Website: EGI—Towards a Sustainable Production Grid Infrastructure. http://www.eu-egi.eu
56. Website: Open European Network for High-Performance Computing on Complex Environments. http://w3.cost.esf.org/index.php?id=177&action_number=IC0805
57. Website: SETI@home. http://setiathome.ssl.berkeley.edu
58. Website: Top500 Supercomputer Sites. http://www.top500.org; Latest Update (2009)
59. Wojick, D., Warnick, W., Carroll, B., Crowe, J.: The digital road to scientific knowledge diffusion: a faster, better way to scientific progress? D-Lib Mag. **12**(6) (2006)
60. Wrzesińska, G., Maassen, J., Bal, H.: Self-adaptive applications on the grid. In: Proceedings of the 12th ACM SIGPLAN Symposium on Principles and Practice of Parallel Programming (PPoPP'07), San Jose, CA, USA, pp. 121–129 (2007)

# Chapter 9
# Application-Level Interoperability Across Grids and Clouds

**Shantenu Jha, Andre Luckow, Andre Merzky,**
**Miklos Erdely, and Saurabh Sehgal**

**Abstract** Application-level interoperability is defined as the ability of an application to utilize multiple distributed heterogeneous resources. Such interoperability is becoming increasingly important with increasing volumes of data, multiple sources of data as well as resource types. The primary aim of this chapter is to understand different ways in which application-level interoperability can be provided across distributed infrastructure. We achieve this by (i) using the canonical wordcount application, based on an enhanced version of MapReduce that scales-out across clusters, clouds, and HPC resources, (ii) establishing how SAGA enables the execution of wordcount application using MapReduce and other programming models such as Sphere concurrently, and (iii) demonstrating the scale-out of ensemble-based biomolecular simulations across multiple resources. We show user-level control of the relative placement of compute and data and also provide simple performance measures and analysis of SAGA–MapReduce when using multiple, different, heterogeneous infrastructures concurrently for the same problem instance. Finally, we discuss Azure and some of the system-level abstractions that it provides and show how it is used to support ensemble-based biomolecular simulations.

S. Jha (✉) · A. Luckow · A. Merzky · S. Sehgal
Louisiana State University, Baton Rouge, 70803, USA
e-mail: sjha@cct.lsu.edu

A. Luckow
e-mail: aluckow@cct.lsu.edu

A. Merzky
e-mail: andre@merzky.net

M. Erdely
University of Pannonia, Veszprem, Hungary
e-mail: erdelyim@gmail.com

M. Cafaro, G. Aloisio (eds.), *Grids, Clouds and Virtualization,*
Computer Communications and Networks,
DOI 10.1007/978-0-85729-049-6_9, © Springer-Verlag London Limited 2011

## 9.1 Introduction

There are numerous scientific applications that either currently utilize or need to utilize data and resources distributed over vast heterogeneous infrastructures and networks with varying speeds and characteristics. Many distributed frameworks are, however, designed with infrastructures; dependence and tight-coupling to specific resource types and technology, in a heterogeneous distributed environment, is not an optimal design choice. In order to leverage the flexibility of distributed systems and to gain maximum runtime performance, applications must shed their dependence on single infrastructure for all of their computational and data processing needs. For example, the Sector/Sphere data cloud is exclusively designed to support data-intensive computing on high-speed networks, while other distributed file systems like GFS/HDFS assume limited bandwidth among infrastructure nodes [11, 20]. Thus, for applications to efficiently utilize heterogeneous environments, abstractions must be developed for the efficient utilization of and orchestration across such distinct distributed infrastructure.

In addition to issues of performance and scale addressed in the previous paragraph, the transition of existing distributed programming models and applications to emerging and novel distributed infrastructure must be as seamless and as nondisruptive as possible. A fundamental question at the heart of all these considerations is the question of how scientific applications can be developed so as to utilize as broad a range of distributed systems as possible, without vendor lock-in, yet with the flexibility and performance that scientific applications demand.

We define Application Level Interoperability (ALI) as a feature that arises, when other than say compiling, there are no further changes required of the application to utilize a new platform. If service-level interoperability can be considered as weak interoperability, ALI can be considered to be *strong* interoperability. The complexity of providing ALI varies and depends upon the application under consideration. For example, it is somewhat easier for simple "distribution-unaware" applications to utilize multiple heterogeneous distributed environments than for applications where multiple distinct and possibly distributed components need to coordinate and communicate.

**The Case for Application-Level Interoperability**   ALI is not only of theoretical interest. There exist many applications which involve large volumes of data on distributed heterogeneous resources. For example, the Earth-Science Grid involves peta- to exa-bytes of data, and one thus cannot move all data (given current transfer capabilities), nor compute at a centralized location. Thus there is an imperative to operate on the data in situ, which in turn involves computation across heterogeneous distributed platforms as part of the same application.

In addition, there exist a wide range of applications that have decomposable but heterogeneous computational tasks. It is conceivable that some of these tasks are better suited for traditional grids, whilst some are better placed in cloud environments. The LEAD application, as part of the VGrADS project provides a prominent example.[1] Due to different data-compute affinity requirement amongst the tasks, some workers might be better placed on a cloud [1], whilst some may optimally be located on regular grids. Complex dependencies and interrelationships between subtasks make this often difficult to determine before runtime.

Last, but not least, in the rapidly evolving world of clouds, there is as of yet little business motivation for cloud providers to define, implement, and support new/standard interfaces. Consequently, there is a case to be made that by providing ALI, such barriers can be overcome, and cross-cloud applications can be easily achieved.

Ideally, an application can utilize any PM, and any PM should be executable on any underlying infrastructure. However, an application maybe better suited to a specific PM; similarly, a specific PM maybe optimized for a specific infrastructure. But where this is not necessarily the case, or more importantly, where/when different application or PM can utilize "nonnative" infrastructure, mix and match across the layers of application, PM and infrastructure should be supported. Currently, many programming models and abstractions are tied to a specific back-end infrastructure. For example, Google's MapReduce, which is tied to Google's file system, or Sphere [21], which is linked to the Sector file system. Thus there is a need to investigate interoperability of different programming models for the same application on different systems.

In our effort to understand these issues, to establish and investigate ALI, we will work with MapReduce and an application based on MapReduce, the canonical wordcount application. We use SAGA, "Simple API for Grid Applications" (see Sect. 9.2), as the programming system. In Ref. [30], for example, we implemented a simple MapReduce-based wordcount application using SAGA. We demonstrated that the SAGA-based implementation is infrastructure independent, whilst still providing control over deployment, distribution, and runtime decomposition. We demonstrated that SAGA–MapReduce is interoperable on traditional (grids) and emerging (clouds) distributed infrastructure concurrently and cooperatively toward a solution of the same problem instance. Our approach was to use the same instance of the wordcounting problem, by using different worker distributions over clouds and grid systems.

A primary focus of this chapter is to build upon and use SAGA-based MapReduce as an exemplar to discuss multiple levels and types of interoperability that can exist between grids and clouds. We will also show that our approach to ALI helps break the coupling between programming models and infrastructure on the one hand, whilst providing empirically driven insight about the performance of an application with different programming models.

This chapter is structured as follows: Sect. 9.2 gives a short overview over those SAGA extensions which enable specifically the ALI work discussed in this paper. Sect. 9.3 describes our SAGA–MapReduce implementation. Section 9.4 discusses the different levels of ALI we are demonstrating in this paper, with more details on the experiments in Sect. 9.5. We change the focus to a more traditional compute-intensive applications in Sect. 9.6; although very commonly used and needed, it

---

[1] http://vgrads.rice.edu/presentations/VGrADS_overview_SC08pdf.pdf

still remains a challenge to effectively utilize multiple resources for the effective solution of ensemble-based simulations. This provides the motivation for our work in
Sect. 9.6. In Sect. 9.7, we discuss Azure in the context of ensemble-based biomolecular simulations. Section 9.8 concludes the chapter with a discussion of the results.

## 9.2 SAGA

The SAGA [23, 32] programming system provides a high-level API that forms a
simple, standard, and uniform interface for the most commonly required distributed
functionality. SAGA can be used to program distributed applications [2, 3] or tool-
kits to manage distributed applications [27], as well as implement abstractions that
support commonly occurring programming, access, and usage patterns.

Figure 9.1 provides an overview of the SAGA programming system's architecture. The SAGA API covers job submission, file access and transfer, and logical file
management. Additionally there is support for Checkpoint and Recovery (CPR),
Service Discovery (SD), and other areas. The API is implemented in C++ and Java,
with Python supported as a wrapper. *saga_core* is the main library, which provides
dynamic support for runtime environment decision making through loading relevant adaptors. We will not discuss SAGA internals here; details can be found else-
where [4, 23].

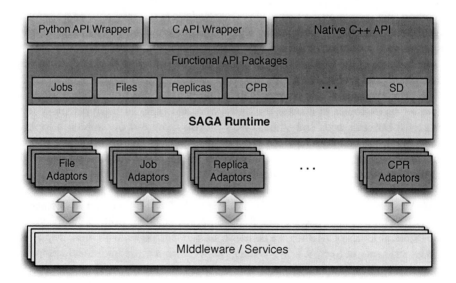

**Fig. 9.1** The SAGA runtime engine dynamically dispatches high-level API calls to a variety of
middlewares

## 9.2.1 Interfacing SAGA to Grids and Clouds

SAGA was originally developed primarily for compute-intensive grids. Ref. [30] demonstrated that in spite of its original design constraints, SAGA can be used to develop data-intensive applications in diverse distributed environments, including clouds. This is in part due to the fact that, at least on application level, much of the "distributed functionality" required for data-intensive applications remains the same. How the respective functionality for grid systems and for EC2-based cloud environments is provided in SAGA is also documented in [30]. Based on those experiences, we added another backend to the set, which allows one to extend the range of backend architectures available to SAGA-C++ to Sector–Sphere [21].

### Sector–Sphere Adaptors: Design and Implementation

Sector and Sphere is a cloud framework specifically designed for writing applications able to utilize the stream processing paradigm. Sector is a distributed file system that manages data across physical compute nodes at the file level and provides the infrastructure to manipulate data in the cloud. Sphere, on the other hand, provides the framework to utilize the stream processing paradigm for processing the data residing on Sector. The Sphere system is composed of Sphere Processing Engines (SPEs) running on the same physical nodes as the Sector file system.

Applications that utilize the stream processing paradigm define a single common function (aka kernel) that is applied to segments of a given data set. When the application invokes Sphere to process data on Sector, the Sphere system retrieves the stream of data, segments the data, and assigns chunks of these segments to the available SPEs for processing.

Sphere allows the user to encode the kernel function in a dynamically linked library written against the Sphere APIs. The SPEs apply this user-defined function to its assigned segments and write the processing results back to files in Sector. This stream of output files can be retrieved by the user from Sector after the processing is complete.

**SAGA Adaptor Overhead**    We execute a simple experiment to measure the overhead introduced by submitting Sphere jobs and Sector file operations through SAGA. The used sample Sphere kernel function accepts a buffer of text and utilizes the Sphere framework to hash words into Sphere buckets, using the first letter as the key. One gigabyte of text data was uploaded to the Sector file system for this test. Furthermore, traces were implemented in the adaptors to measure the exact time spent in SAGA processing and translation before the raw Sphere APIs were called. As seen in Table 9.1, the SAGA overhead is, if compared to the overall execution time of the application, negligible. This makes SAGA an excellent platform to compare Sphere with other distinct programming models.

**Table 9.1** Adaptor overhead measurements from processing 8 GBs of data with 8 SPEs running the wordcount application on 8 physical nodes on Poseidon (a LONI cluster). All times are in minutes, aggregated from 10 runs

|         | Vanilla Sphere | SAGA–Sphere | Adaptor overhead |
|---------|----------------|-------------|------------------|
| Mean    | 3.2            | 4.1         | 0.43             |
| Stdev   | 0.5            | 1.2         | 0.07             |

Thanks to the low overhead of developing SAGA adaptors, we were able to implement the Sector file adaptor and the Sphere job adaptor for applications to utilize the stream processing paradigm through SAGA. The enhancement of SAGA–MapReduce, along with the implementation of the Sector/Sphere adaptors naturally gives us the opportunity to compare and study these two distinct programming models.

## 9.3 SAGA-Based MapReduce

Given its relevance [30], we choose the SAGA–MapReduce implementation to compare both, different backend systems (grids, clouds, and clusters) and different programming models (master/slave, Sector–Sphere streams). A simple wordcount application on top of SAGA–MapReduce has been used as a close-to-reality test case and is described in Sect. 9.5.

### 9.3.1 SAGA–MapReduce Implementation

Our implementation of SAGA–MapReduce interleaves the core MapReduce logic with explicit instructions on where processes are to be scheduled. The advantage of this approach is that our implementation is no longer bound to run on a system providing the appropriate semantics originally required by MapReduce and is portable to a broader range of generic systems as well. The drawback is that it is more complicated to extract performance, as some system level semantics has to be recreated in application space (i.e., on SAGA or SAGA–MapReduce) level. The fact that the implementation is single threaded proved to be the primary current performance inhibitor. However, none of these complexities are exposed to the end-user, as they remain hidden within the framework.

SAGA–MapReduce exposes a simple interface which provides the complete functionality needed by any MapReduce algorithm, while hiding the more complex functionality, such as chunking of the input, sorting the intermediate results, launching and coordinating the workers, etc.—these are generically implemented by the framework. The application consists of two independent processes, a master and worker processes. The master process is responsible for:

- launching all workers for the map and reduce steps, as described in a configuration file provided by the user; and
- coordinating the workers, chunking of the data, assigning the input data to the workers of the map step, handling the intermediate data files produced by the map step, passing the location of the sorted output files to the workers of the reduce step.

When launching a job, the master is the executable run by the client itself, which means that the resource from which the client program is run determines from where the master will be available. A MapReduce job is specified by a JobDescription object in which the user sets the Mapper and Reducer classes, input and output paths and formats. The used InputFormat creates the logical partitions of the input data for the master which information is then sent to idle workers. A RawRecordReader implementation is responsible for interpreting an InputChunk and providing a record iterator for the Mapper. It is possible to support any kind of data source for which a record-oriented view makes sense by writing a custom RawRecordReader. The output from the Mapper is further processed by the Partitioner which assigns emitted key/value pairs from the Mapper to reducers. Finally, a RawRecordWriter writes output data to files. Custom RawRecordWriter and Partitioner classes can be also implemented to suit the application's needs.

The master process is readily available to the user and needs no modification for different map and reduce functions to execute. The worker processes get assigned work either from the map or the reduce step. The functionality for the different steps have to be provided by the user, which means that the user has to write two C++ functions implementing the respective MapReduce kernels.

Both the master and the worker processes use the SAGA-API as an abstract interface to the used infrastructure, making the application portable between different architectures and systems. The worker processes are launched using the SAGA job package, allowing the jobs to launch either locally, on Globus/GRAM backends, on EC2 instances, through SSH or on a Condor pool. The communication between the master and workers is ensured by using the SAGA advert package, abstracting an information database in a platform-independent way, and the SAGA stream package, abstracting streaming data access between network endpoints. The master creates logical partitions of the data (referred to as chunking, analogous to Google's MapReduce), so the data-set does not have to be split and distributed manually. The input data can be located on any file system supported by SAGA, such as the local file system or a distributed file system like HDFS or KFS [15].

## 9.3.2 Enhancing SAGA-Based MapReduce Performance

The performance enhancements to the SAGA–MapReduce implementation as discussed in [30] are based on two important changes: (i) rearranging the shuffle phase and (ii) using a serialized binary format instead of plain text for intermediate data storage (also available as an input and output format). The first change means that, instead of having the master merge and then sorting the intermediate data by key before entering the reduce phase, the workers buffer key/value pairs from the map phase and store them in sorted order on disk, doing an in-memory sort before writing. Also, since intermediate key/value pairs from a map worker are already sorted, the reduce workers need to only merge these pairs coming from different map workers while applying the user-defined reduce function to the merged intermediate key and value list. The second enhancement applies to the storage of the intermediate key/value pairs in a so-called *sequence file format*. This file format allows storing of serialized key/value objects which can be read and merged much faster in the reduce phase than text data, as there is no need for costly parsing. We used the Google Protocol Buffers library for implementing serialization [5]. The processing of input and output key/value pairs is further enhanced by minimizing unnecessary memory I/O operations using a zero-copy scheme.

### 9.3.3 SAGA–MapReduce Set-Up

As with any application which concurrently spans multiple diverse resources or infrastructures, the coordination between the different application components becomes challenging. The SAGA–MapReduce implementation uses the SAGA advert API for that task and can thus limit the a priori information needed for bootstrapping the application: the compute clients (workers) require (i) the contact address of the used advert service instance and (ii) a unique worker ID to register with in that advert service, so that the master can start to assign work items. Both information are provided by the master via command line parameters to the worker, at startup time.

The master application requires the following additional information: (i) a set of resources where the workers can execute, (ii) the location of the input data, (iii) the target location for the output data, and (iv) the contact point for the advert service for coordination and communication.

In a typical configuration, for example, three worker instances could be started: the first could be started via GRAM and PBS on qb.teragrid.org, the second on a pre-instantiated EC2 image (instance-id i-760c8c1f), and the third on a dynamically deployed EC2 instance (no instance id given). Note that the startup times for the individual workers may vary over several orders of magnitudes, depending on the PBS queue waiting time and VM startup time. The MapReduce master will start to utilize workers as soon as they are able to register themselves and so will not wait until all workers are available. That mechanism both minimizes time-to-solution and maximizes resilience against worker loss. A simple parameter controls the number of workers created on each compute node; as we will see by varying this parameter, the chances are good that compute and communication times can be interleaved and that the overall system utilization can increase (especially in the absence of precise knowledge of the execution system).

## 9.4 Application Level Interoperability: Three-Levels

The motivation of ALI across multiple, heterogeneous, and distributed resources follows from large-scale scientific applications, such as the Earth Science Grid and LEAD. However for simplicity of treatment and to focus on the levels of interoperability, we will use a simple, self-contained application that has also become the *canonical* MapReduce application driver, wordcount.

### 9.4.1 Interoperability Types

Using the wordcount application, we will demonstrate three types of application level interoperability. We outline them here:

**Type I: Application Interoperability via Adaptors**

As discussed, SAGA provides the ability to load a wide range of system-specific adaptors dynamically. Thus a simple form of interoperability, possibly specific to applications developed using SAGA, is that an application can use any distributed systems without changes to the application, thus experiencing cloud–cloud or grid–cloud interoperability. We refer to this as Type I interoperabilty.

Thanks to the relative simplicity of developing SAGA adaptors, SAGA has been successfully interfaced to three cloud systems: Amazon's EC2, Eucalyptus [18] (a local installation of Eucalyptus at LSU), and Nimbus [6]; and also to a multitude of grid-based environments, including TeraGrid, Loni, and NGS. SAGA-based applications are thus inherently able to utilize this form of ALI.

**Type II: Application Interoperability Using Programming Models**

Interoperability at a higher level than adaptors is both possible and often desirable. An application can be considered interoperable if it is able to switch between back-end specific programming models. We will discuss an example where the wordcount application is implemented so that it can utilize either a Sector–Sphere framework via SAGA for the OCC backend, or the SAGA–MapReduce framework for generic grid and cloud backends.

**Type III: Application Interoperability Using Different Programming Models for Concurrent Execution**

At another level, an application can also be considered interoperable when it executes multiple programming models *concurrently* over diverse backends. We demonstrate that a wordcount application uses both Sector–Sphere and SAGA–MapReduce when spanning multiple backends. The challenges of having different parts of an application execute concurrently; using different programming models is conceptually different to loading different adaptors concurrently. Thus we describe this as a separate type of interoperability.

### 9.4.2 Experimental Setup

Simulations were performed on shared TeraGrid-LONI (Louisiana Optical Network Initiative) [7] resources running Globus and SSH; on GumboGrid, a small cluster at LSU running Eucalyptus; on Amazon's EC2; on a bare 50-node cluster of the Hungarian Academy of Sciences; and on the OCC testbed [8] running Sector–Sphere. Jobs are started via the respectively available middlewares, via SAGA's job API. Data exchange is either performed via streams or via SAGA's file transfer API, which can dynamically switch between the various available protocols.

For cloud environments, we support the runtime configuration of VM instances by staging a preparation script to the VM after its creation and executing it with root permissions. In particular for apt-get Linux distribution, the post-instantiation software deployment is actually fairly painless but naturally adds a significant amount of time to the overall VM startup (which encourages the use of preconfigured images).

For experiments in this paper, we prepared custom VM images with preinstalled prerequisites. We utilize preparation scripts solely for some fine tuning of parameters: for example, to deploy custom saga.ini files or to ensure the finalization of service startups before application deployment.

Deploying SAGA–MapReduce framework and the wordcount application on different grids, clouds or clusters requires adapting the configuration to the specific environment. For example, when running SAGA–MapReduce on EC2, the master process resides on one VM, while workers reside on different VMs. Depending on the available adaptors, Master and Worker can perform either local I/O on a global/distributed file system or remote I/O on a remote, nonshared file system.

It must be noted that we utilized different SAGA–MapReduce versions for the described experiments: the work described in this paper spans more than 18 months, and the SAGA–MapReduce implementation has simply evolved over time. As our primary goal is to demonstrate interoperability and not to document maximal performance, we consider those results valid nonetheless.

## 9.5 Interoperability Experiments: Wordcount

We use our own implementations of the well-known wordcount application for our experiments. Wordcount has a well-understood runtime and scaling behavior, and thus serves us well for focusing the tests on the used frameworks and middlewares.

The MapReduce based wordcount implementation is described in [30]. For the Sector–Sphere version of wordcount, we implemented two kernel functions. The first one is responsible for hashing the words in the data set into different "buckets," depending on the word's starting letter. The standard C++ collate hashing function was used for this purpose. The second kernel function reads each hash bucket, sorts the words in memory, and outputs the final count of the words in the data set. For example, a file containing the words (`'bread'` `'bee'` `'bee'` `'honey'`) would be hashed into buckets as (`'bread'` `'bee'` `'bee'`) and (`'honey'`). The second kernel function would read these intermediate bucket files, sort the words, and produce the result (`.bread 1.`, `.bee 2.`, `.honey 1.`). The Sphere system is responsible for assigning files for processing, synchronization, and writing output results back to Sector.

### 9.5.1 Type I ALI: Interoperability via Adaptors

In an earlier paper (Ref. [30]), we performed tests to demonstrate how SAGA–MapReduce utilizes different infrastructures and provides control over task-data placement; this led to insight into performance on "vanilla" grids. This work extends earlier work and establishes that SAGA–MapReduce can provide cloud–cloud interoperability and cloud–grid interoperability. We performed the following experiments:

1. We compare the performance of SAGA–MapReduce when exclusively running on a cloud platform to that when on grids. We vary the number of workers (1 to 10) and the data-set sizes varying from 10 MB to 1 GB.
2. For clouds, we then vary the number of workers per VM, so that the ratio is 1:2 and 1:4, respectively.
3. We then distribute the same number of workers across two different clouds, EC2 and Eucalyptus.
4. Finally, for a single master, we distribute workers across grids (QueenBee on the TeraGrid) and clouds (EC2 and Eucalyptus) with one job per VM.

It is worth reiterating that although we have captured concrete performance figures, it is not the aim of this work to analyze the data and provide a performance model. In fact, it is difficult to understand performance implications, as a detailed analysis of the data and understanding the performance will involve the generation of "system probes," as there are differences in the specific cloud system implementation and deployment. In a nutshell, without adjusting for different system implementations, it is difficult to rigorously compare performance figures for different configurations on different machines. At best we can currently derive trends and qualitative information. Any further analysis is considered out of scope for this paper.

It takes SAGA about 45 s to instantiate a VM on Eucalyptus and about 200 s on average on EC2. We find that the size of the image (say 5 GB versus 10 GB)

influences the time to instantiate an image but is within image-to-image instantiation time fluctuation. Once instantiated, it takes from 1–10 s to assign a job to
an existing VM on Eucalyptus or EC2. The option to tie the VM lifetime to the
saga::job_service object lifetime is a configurable option. It is also a matter
of simple configuration to vary how many jobs (in this case workers) are assigned
to a single VM: the default is 1 worker per VM. The ability to vary this number is
important—as details of actual VMs can differ—and useful for our experiments.

### Results and Analysis

The total time-to-solution ($T_s$) of a SAGA–MapReduce job can be decomposed as
the sum of three primary components $t_{pre}$, $t_{comp}$, and $t_{coord}$. Here $t_{pre}$ is defined as
preprocessing time, which covers the time to chunk the data into fixed-size data
units, to distribute them, and also to spawn the job. $t_{pre}$ does not include the time
required to start VM instances. $t_{comp}$ is the time to actually compute the map and reduce function on a given worker, whilst $t_{coord}$ is the time taken to assign the payload
to a worker, update records, and to possibly move workers to a destination resource;
in general, $t_{coord}$ scales as the number of workers increases.

Table 9.2 shows performance measurements for a variety of worker placement
configurations. The master places the workers on either clouds or on the TeraGrid

**Table 9.2** Performance data for different configurations of worker placements

| #workers | | Data size (MB) | $T_s$ (s) | $T_{sp}$ (s) | $T_s - T_{sp}$ (s) |
|---|---|---|---|---|---|
| TG | AWS | | | | |
| 4 | – | 10 | 8.8 | 6.8 | 2.0 |
| – | 1 | 10 | 4.3 | 2.8 | 1.5 |
| – | 2 | 10 | 7.8 | 5.3 | 2.5 |
| – | 3 | 10 | 8.7 | 7.7 | 1.0 |
| – | 4 | 10 | 13.0 | 10.3 | 2.7 |
| – | 4 (1) | 10 | 11.3 | 8.6 | 2.7 |
| – | 4 (2) | 10 | 11.6 | 9.5 | 2.1 |
| – | 2 | 100 | 7.9 | 5.3 | 2.6 |
| – | 4 | 100 | 12.4 | 9.2 | 3.2 |
| – | 10 | 100 | 29.0 | 25.1 | 3.9 |
| – | 4 (1) | 100 | 16.2 | 8.7 | 7.5 |
| – | 4 (2) | 100 | 12.3 | 8.5 | 3.8 |
| – | 6 (3) | 100 | 18.7 | 13.5 | 5.2 |
| – | 8 (1) | 100 | 31.1 | 18.3 | 12.8 |
| – | 8 (2) | 100 | 27.9 | 19.8 | 8.1 |
| – | 8 (4) | 100 | 27.4 | 19.9 | 7.5 |

**Table 9.3** Performance data for different configurations of worker placements on TG, Eucalyptus-Cloud, and EC2

| #workers | | | Data size (MB) | $T_s$ (s) | $T_{sp}$ (s) | $T_s - T_{sp}$ (s) |
|---|---|---|---|---|---|---|
| TG | AWS | Eucal. | | | | |
| – | 1 | 1 | 10 | 5.3 | 3.8 | 1.5 |
| – | 2 | 2 | 10 | 10.7 | 8.8 | 1.9 |
| – | 1 | 1 | 100 | 6.7 | 3.8 | 2.9 |
| – | 2 | 2 | 100 | 10.3 | 7.3 | 3.0 |
| 1 | – | 1 | 10 | 4.7 | 3.3 | 1.4 |
| 1 | – | 1 | 100 | 6.4 | 3.4 | 3.0 |
| 2 | 2 | – | 10 | 7.4 | 5.9 | 1.5 |
| 3 | 3 | – | 10 | 11.6 | 10.3 | 1.6 |
| 4 | 4 | – | 10 | 13.7 | 11.6 | 2.1 |
| 5 | 5 | – | 10 | 33.2 | 29.4 | 3.8 |
| 10 | 10 | – | 10 | 32.2 | 28.8 | 2.4 |

(TG). The configurations, separated by horizontal lines, are classified as either all workers on the TG or having all workers on EC2. For the latter, unless otherwise indicated in parentheses, every worker is assigned to a unique VM. In the final set of rows, the number in parentheses indicates the number of VMs used. Note that the spawning times depend on the number of VMs, even if it does not include the VM startup times.

Table 9.3 shows data from our interoperability tests. The first set of data establishes cloud–cloud interoperability. The second set (rows 5–11) shows interoperability between grids–clouds (EC2). The experimental conditions and measurements are similar to Table 9.1.

We find that in our experiments $t_{comp}$ is typically greater than $t_{coord}$, but when the number of workers gets large, and/or the computational load per worker small, $t_{coord}$ can dominate (internet-scale communication) and increase faster than $t_{comp}$ decreases; thus, overall $T_s$ can increase for the same dataset size, even though the number of independent workers increases. The number of workers associated with a VM also influences the performance, as well as the time to spawn; for example, as shown by the three lower boldface entries in Table 9.1, although four identical workers are used depending upon the number of VMs used, $T_c$ (defined as $T_S - T_{spawn}$) can be different. In this case, when four workers are spread across four VMs (i.e., default case), $T_c$ is lowest, even though $T_{spawn}$ is the highest; $T_c$ is highest when all four are clustered onto one VM. When exactly the same experiment is performed using dataset of size 10 MB, it is interesting to observe that $T_c$ is the same for four workers distributed over one VM as it is for four VMs, whilst when the performance for the case where four workers are spread-over two VMs out-perform both (2.1 s).

Table 9.3 shows performance figures when equal numbers of workers are spread across two different systems; for the first set of rows, workers are distributed on

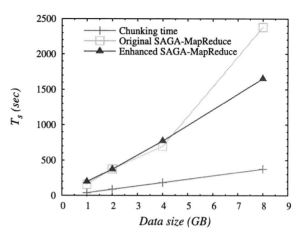

**Fig. 9.2** Comparison of enhanced SAGA–MR performance versus early version of SAGA–MR on the ILAB cluster using eight workers running on eight physical machines. Jobs were launched via SSH and used NFS for file operations

EC2 and Eucalyptus. For the next set of rows, workers are distributed over the TG and Eucalyptus, and in the final set of rows, workers are distributed between the TG and EC2. Given the ability to distribute at will, we compare performance for the following scenarios: (i) when four workers are distributed equally (i.e., two on each) on a TG machine and on EC2 (1.5 s) with the scenarios when (ii) all four workers are either exclusively on EC2 (2.7 s) or (iii) all workers are on the TG machine (2.0 s) (see Table 9.1, boldface entries on the first and fifth lines). It is *interesting* that, in this case, $T_c$ is lower in the distributed case than when all workers are executed locally on either EC2 or TG; we urge that not too much be read into this, as it is just a coincidence that a *sweet spot* was found where on EC2, four workers had a large spawning overhead compared to spawning two workers, and an increase was in place for two workers on the TG. Also it is worth reiterating that for the same configuration, there are experiment-to-experiment fluctuations (typically less than 1 s). The ability to enhance performance by distributed (heterogeneous) workloads across different systems remains a distinct possibility; however, we believe more systematic studies are required.

The original SAGA–MapReduce version (as used for the experiments presented above) physically chunked the input data files. Our evolved version, however, creates logical chunks (i.e., no file writing takes place). It is thus fair to compare their time-to-solution performance by subtracting the chunking time from the early version's job completion time. Figure 9.2 thus shows the corrected performance data for the early version and enhanced version of SAGA–MapReduce. Eight workers were spawned via the SAGA SSH adaptor on eight physical machines, and data were exchanged through a shared NFS file system. The figure shows that the SAGA–MapReduce enhancements make a difference for larger data sets. This can be attributed to the fact that the more efficient shuffle phase implementation, which reduces disk I/O and CPU usage in the reduce phase by doing only a merge, outperforms the old implementation, which performed a merge–sort of all the intermediate output files.

### 9.5.2 Type II ALI: Application Performance Using SAGA-Based Sphere and MapReduce

**Experiment I: Varying Chunk Sizes**

For the Sector–Sphere-based wordcount, Sector maintains and tracks data in the cloud at the file level. There is no support in Sphere to control the chunk sizes of files assigned to the Sphere processing engines. Therefore, to experiment with different chunk sizes, the files were split manually into smaller chunks before the wordcount application was launched. In this set of experiments, we vary the chunk size from 16 MB to 256 MB, while keeping the number of SPEs constant at 8, and the data size constant at 4 GB. Each SPE is running on a separate physical node in the cluster. These results are presented in Fig. 9.3. Note that both data and computation were distributed for these experiments. As evident from the results we collected, a correlation exists between the chunk sizes and performance of Sphere. As the chunk sizes increase, the performance deteriorates. In particular, we observe a decline in performance after the 64-MB data, in that there is an increase in the gradient of the plot, after chunk sizes have reached 64 MB.

We performed the same set of experiments with SAGA–MapReduce-based word-count and observed a completely different performance trend as the chunk size varied. We use an HDFS file system running datanodes on each of the eight workers and set the number of reduce tasks to 8. In case of SAGA–MapReduce, performance increases with larger chunk sizes, reaching Sphere's performance at the 256-MB data point. According to our analysis, this can be attributed to the fact that for SAGA–MapReduce, $t_{coord}$ can dominate the total time-to-solution for large number of workers (or equivalently smaller chunk sizes). The larger the chunk size, the smaller the number of map tasks launched (workers), and provided that the data workload assigned to each worker is not too high, $t_{coord}$ decreases with increasing chunk size.

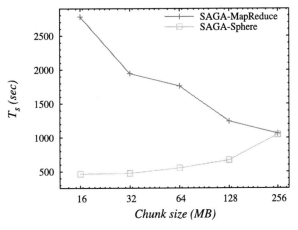

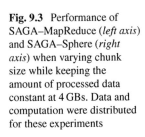

**Fig. 9.3** Performance of SAGA–MapReduce (*left axis*) and SAGA–Sphere (*right axis*) when varying chunk size while keeping the amount of processed data constant at 4 GBs. Data and computation were distributed for these experiments

## Experiment II: Varying Workers

We perform two sets of experiments with SAGA–Sphere running the same word-count application. We keep the chunk size constant at 64 MB, the data size fixed at 4 GB (as previously), but vary the number of workers in two configurations. In the first configuration, we use Sphere on a local data and local compute configuration. In the second configuration, we observe how the solution scales to a distributed data and distributed compute configuration. These results are illustrated in Fig. 9.4.

For the local–local configuration, we launch Sector and Sphere on a single physical node. For the distribute–distribute configuration, we launch Sector and Sphere on one physical node per SPE. For the dataset sizes considered, we observe good performance from four to six distributed workers, before which the coordination costs due to the number of SPEs starts to get high, with a concomitant increase in time-to-solution. This is a nice but simple demonstration of the advantage of distribution (logically distributed in this case, if not physically distributed). Sector can maintain file replicas to achieve optimal data distribution between SPEs and minimize synchronization overhead. For the purpose of our experiments, we limited Sector to not create any replicas.

We perform similar measurements via SAGA–MapReduce: while keeping the chunk size constant at 64 MB and the data size at 4 GB, we note the time-to-solution of the wordcount application in the two configurations described for SAGA–Sphere above. For the distribute–distribute configuration, we use HDFS as the distributed file system and launch jobs using the SAGA SSH job adaptor. In contrast to the varying chunk size experiments, we set the number of reduce tasks to be equal to the number of workers spawned. For the local–local configuration, we launch workers on the machine running the master as separate processes and use the local file system for file operations. As can be seen in Fig. 9.5, SAGA–MapReduce scales as expected in the distribute–distribute configuration. However, for the local–local configuration, performance degrades when adding more workers. Since each map task writes as many files as the number of reduce tasks at the same time, and each

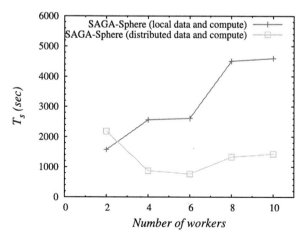

**Fig. 9.4** Comparison of SAGA–Sphere performance when varying the number of workers between 2 and 10 in two configurations: (1) local data and computation and (2) distributed data and computation

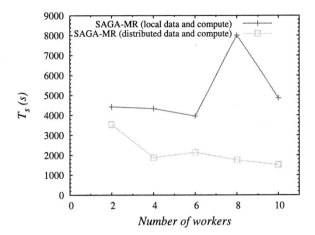

**Fig. 9.5** Comparison of SAGA–MapReduce when varying the number of workers between 2 and 10 in two configurations: (1) local data and computation and (2) distributed data and computation

reduce task needs to read from as many files as the number of input chunks, the number of concurrent disk I/O increases very quickly; this can cause a bottleneck when performing computations on one physical node (I/O system). According to Fig. 9.5, the optimal number of workers is 6 for the local–local configuration. We did not include results for more than 10 workers in order to preserve the linear scale.

SAGA gives us the opportunity to experiment with different programming models very easily. As evident from the data plots in Fig. 9.3, certain behavioral trends for SAGA–Sphere and SAGA–MapReduce emerge. In Experiment 1, where we keep the number of workers constant and vary the chunk sizes, the trends between SAGA–MapReduce and Sphere are inversed: the Sphere performance deteriorates with increasing chunk sizes, while the performance of SAGA–MapReduce increases. This behavior suggests that SAGA–MapReduce's synchronization overhead to manage smaller chunk sizes compared to the speed up achieved through parallelism is much higher. In the case of the wordcount application, SAGA–MapReduce appears to be more suitable for coarse grained computations. SAGA–Sphere, on the other hand, yields better performance from smaller chunks sizes (a larger amount of files), making it suitable for finer grained computations with better data distribution.

In Experiment 2, where we keep the chunk size to a constant of 64 MB, SAGA–Sphere exhibits a trend where adding more SPEs has a positive impact on performance. However, at the 8 SPEs and 10 SPEs data points, we see a decline in performance, possibly due to high synchronization costs between the workers. What is interesting to notice are the two data points at 64-MB chunk size and at 16-MB chunk size for SAGA–Sphere in Fig. 9.3. Reducing the chunk size by 75%, thereby providing better data distribution with a larger number of files, we see almost a 65% increase in performance. This further confirms our supposition that good data distribution had a major impact on Sphere's performance for the wordcount application. We do not claim that fine grained computation granularity is a concrete determinant of SAGA–Sphere's performance in a general case, but a noticeable aspect that emerges through its comparison with SAGA–MapReduce.

### 9.5.3 Type III ALI: Interoperability Concurrency

We discuss the third type of interoperability in this section, where SAGA–MapReduce and SAGA–Sphere are used in conjunction to solve the wordcount problem. We first use the "netperf" utility to measure the throughput from the client host to the SAGA–MapReduce master node and SAGA–Sphere master node. In our case, the throughput measured to the two nodes was approximately equal (935 MB/s to SAGA–Sphere and 925 MB/s to SAGA–MapReduce). Based on these metrics, we split the 4.0-GB data set into two equal 2.0-GB parts. This is to ensure that the data transfer time to both masters is approximately the same. We configured both systems to utilize four workers each and 64-MB chunk sizes. The data transfer time to the Sector cloud took a total of 97.8 seconds and 10.4 seconds to the SAGA–MapReduce master node. The longer transfer time to Sector can be credited to the overhead incurred from registering the files in Sector. The data transfer was done sequentially.

We found that SAGA–Sphere took a total of 441.3 seconds to process the 2.0-GB data, while SAGA–MapReduce took a total of 769 seconds. Aggregating the output results from the two systems took a negligible amount of time (only 0.9 seconds). The data was already sorted and hence could be merged in almost constant time. The above simple experiment of combining two varied programming models for solving a common problem paves the way to further investigation into smarter data and compute placement techniques. The total time taken to execute the wordcount application in this case was approximately 877.9 seconds. It is interesting to note that this performance measure lies between 1329.96 seconds for 8 SPEs and 716 seconds for 8 SAGA–MapReduce workers at a 64-MB chunk size.

## 9.6 Interoperability Experiments: Ensemble of Biomolecular Simulations

Several classes of applications are well suited for distributed environments. Probably the best known and most powerful examples are those that involve an ensemble of decoupled tasks, such as simple parameter sweep applications [12]. In the following we investigate an ensemble of (parallel HPC) MD simulations. Ensemble-based approaches represent an important and promising attempt to overcome the general limitations of insufficient time-scales and specific limitations of inadequate conformational sampling arising from kinetic trappings. The fact that one single long-running simulation can be substituted for an ensemble of simulations, make these ideal candidates for distributed environments. This provides an important general motivation for researching ways to support scale-out and thus enhance sampling and to thereby increase "effective" time-scales studied.

The physical system we investigate is the HCV internal ribosome entry site and is recognized specifically by the small ribosomal subunit and eukaryotic initiation factor 3 (eIF3) before viral translation initiation. This makes it a good candidate for new drugs targeting HCV. The initial conformation of the RNA is taken from the

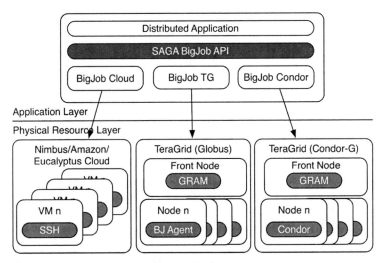

**Fig. 9.6** An overview of the SAGA-based Pilot Job: The SAGA Pilot Job API is currently implemented by three different backends: one for grids, Condor, and for clouds

NMR structure (PDB ID: 1PK7). By using multiple replicas, the aim is to enhance the sampling of the conformational flexibility of the molecule and the equilibrium energetics. NAMD [31] is used as MD code.

To efficiently execute the ensemble of batch jobs without the necessity to queue each individual job, the application utilizes the SAGA BigJob framework [28]. BigJob is a Pilot-Job framework that provides the user a uniform abstraction to grids and clouds independent of any particular cloud or grid provider that can be instantiated dynamically. Pilot-Jobs are an execution abstraction that have been used by many communities to increase the predictability and time-to-solution of such applications. Pilot-Jobs have been used to (i) improve the utilization of resources, (ii) to reduce the net wait time of a collection of tasks, (iii) facilitate bulk or high-throughput simulations where multiple jobs need to be submitted which would otherwise saturate the queuing system, and (iv) as a basis to implement application-specific scheduling decisions and policy decisions.

As shown in Fig. 9.6, BigJob currently provides an abstraction to grids, Condor pools, and clouds. Using the same API, applications can dynamically allocate resources via the big-job interface and bind subjobs to these resources.

In the following we use an ensemble of MD simulations to investigate different BigJob usage modes and analyze the time-to-completion $T_c$ in different scenarios.

### 9.6.1 Scenario A: $T_c$ for Workload for Different Resource Configurations

In this scenario and as proof of scale-out capabilities using type I interoperability provided by SAGA, we use SAGA-BigJob (a higher-level package) to run replicas

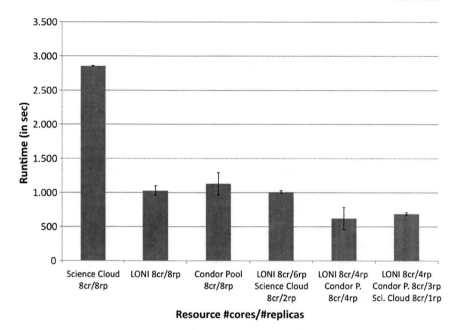

**Fig. 9.7** A1–A3: Collective Usage of Grid, Condor, and Cloud Resources for Workload of 8 Replicas: The experiments showed that if the grid and Condor resource Poseidon has only a light load, no benefits for using additional cloud resources exist. However, the introduction of an additional Condor or grid resource significantly decreases $T_c$

across different types of infrastructures At the beginning of the experiment, a particular set of Pilot-Jobs is started in each environment. Once a Pilot-Job becomes active, the application assigns replicas to this job. We measure $T_c$ for different resource configurations using a workload of eight replicas each running on eight cores. The following setups have been used:

- Scenario 1 (A1): Resource I and III – Clouds and GT2-based grids
- Scenario 2 (A2): Resource II and III – Clouds and Condor grids
- Scenario 3 (A3): Resource I, II and III – Clouds, GT2 and Condor grids

For this experiment, the LONI clusters, Poseidon and Oliver, are used as grid and Condor resources and Nimbus as cloud resource.

Figure 9.7 shows the results. For the first three bars, only one infrastructure was used to complete the 8-replica workload. Running the whole scenario in the Science Cloud resulted in a quite poor but predictable performance—the standard deviation for this scenario is very low. The LONI resources are about three times faster than the Science Cloud, which corresponds to our earlier findings. The performance of the Condor and grid BigJob is similar, which can be expected since the underlying physical LONI resources are the same. Solely, a slightly higher startup overhead can be observed in the Condor runtimes.

In the next set of three experiments, multiple resources were used. For Scenario A1 (the fourth bar from left), two replicas were executed on the Science Cloud.

The offloading of two replicas to an additional cloud resource resulted in a light improvement of $T_c$ compared to using just LONI resources. Thus, the usage of cloud resources must be carefully considered since $T_c$ is determined by the slowest resource, i.e., Nimbus. As described earlier, the startup time for Nimbus images is in particular for such short runs significant. Also, NAMD performs significantly worse in the Nimbus cloud than on Poseidon or Oliver. Since the startup time on Nimbus averages to 357 s and each of eight core replicas runs for about 363 s, at least 720 s must be allowed for running a single replica on Nimbus. Thus, it can be concluded that if resources in the grids or Condor pool are instantly available, it is not reasonable to start additional cloud resources. However, it must be noted that there are virtual machines types with a better performance available, e.g., in the Amazon cloud. These VMs are usually associated with higher costs (up to 2.40 $ per CPU hour) than the Science Cloud VMs. For a further discussion of cost trade-offs for scientific computations in clouds, see Deelman et al. [17].

### 9.6.2 Scenario B: Investigating Workload Distribution for a Given $T_{max}$

Given that clouds provide the illusion of infinite capacity, or at least queue wait-times are nonexistent, it is likely that when using multiple resource types and with loaded grids/clusters (e.g., TeraGrid is currently oversubscribed, and typical queue wait times often exceed 24 hours), most subjobs will end up on the cloud infrastructure. Thus, in Scenario B, the resource assignment algorithm we use is as follows: We submit tasks to noncloud resources first and periodically monitor the progress of the tasks. If insufficient jobs have finished when time equal to $T_X$ has elapsed (determined per criteria outlined below), then we move the workload to utilize clouds. The underlying basis is that clouds have an explicit cost associated with them, and if jobs can be completed on the TG/Condor-pool while preserving the performance constraints, we opt for such a solution. However, if queue loads prevent the performance requirements being met, we move the jobs to a cloud-resource, which we have shown has less fluctuation in $T_c$ of the workload.

For this experiment, we integrated a progress manager that implements the described algorithm into the replica application. The user has the possibility to specify a maximum runtime and a check interval. At the beginning of each check interval, the progress manager compares the number of jobs done with the total number of jobs and estimates the total number of jobs that can be completed within the requested timeframe. If the total number of jobs is higher than this estimate, the progress monitor instantiates another BigJob object request additional cloud resources for a single replica. In this scenario, each time an intermediate target is not met, four additional Nimbus VMs sufficient for running another eight-core replica are instantiated.

In the investigated scenario, we configured a maximum runtime of 45 minutes and a progress check interval of 4 minutes. We repeated the same experiment 10 times at different times of the day. In 6 out of 10 cases the scenario completed in about 8 minutes. However, the fluctuation in particular in the waiting time on typical grid resources can be very high. Thus, in four cases it was necessary to start additional VMs to meet the application deadline. In two cases three Pilot-Jobs, each with eight cores, had to be started, and in one case a single Pilot-Job was sufficient. In a single case the deadline was missed solely due to the fact that not enough cloud resources were available, i.e., we are only able to start two instead of three Pilot-Jobs.

## 9.7 Future Work: Windows Azure

As alluded to in earlier parts of this chapter, we have investigated and analyzed multiple platforms, existing (GT2-based canonical Grids) and emerging infrastructure (EC2). Azure is an emerging cloud platform developed and operated by Microsoft. Azure provides different abstractions, building blocks for creating scalable and reliable scientific applications without the need for on-premise hardware; we believe that Azure-based abstractions and services have the potential to be very effective in the design, development, and deployment of distributed applications. Due to these capabilities, we will focus on Azure in the closing section of this chapter, but as many details are still being worked out, we present this analysis as future (high-potential) work.

Azure follows the platform as a service paradigm offering an integrated solution for managing compute- and data-intensive tasks as well as web applications. The platform is able to dynamically scale applications without the need to manually manage tasks and deployments on virtual machine level. After a brief introduction of the abstractions that are provided of Azure, we discuss how Azure's abstractions and capabilities can be utilized for ensemble-based biomolecular simulations of the type discussed in Sect. 9.6.

### 9.7.1 Understanding Azure-System Abstractions

Azure provides different higher-level services, e.g., the Azure AppFabric or Azure Storage, that can be accessed via HTTP/REST from anywhere. Windows Azure offers a platform for on-demand computing and for hosting generic server-side applications. The so-called Azure fabric controller automatically monitors alls VMs, automatically reacts to hardware and software failures, and manages application upgrades.

#### Compute

Windows Azure formalizes different types of virtual machines into so-called roles. Web roles, e.g., are used to host web applications and frontend code, while worker roles are well suited for background processing. While these roles target specific scenarios, they are also highly customizable. Worker roles can, e.g., run native code. The application must solely implemented a defined entry point, which is then called by Azure. The Azure fabric controller automatically manages and monitors applications, handles hardware and software failures, as well as updates to the operating system or to the application. Commonly, scientific applications utilize worker roles for compute- and data-intensive tasks. AzureBlast [26], e.g., heavily relies on worker roles for computing biosequences.

**Storage**

For storing large amounts of data, the Azure storage platform provides three key services: the *Azure Blob Storage* for storing large objects of raw data, the *Azure Table Storage* for semi-structured data, and the *Azure Queue Storage* for implementing message queues. The data is storage replicated across multiple data centers to protect it against hardware and software failures. In contrast to other cloud offerings (e.g., Amazon S3), the Azure Storage Services provide strong consistency guarantees, i.e., all changes are immediately visible to all future calls. While eventual consistency as implemented by S3 [16] usually offers a better performance and scalability, it has some disadvantages mainly caused by the fact that the complexity is moved to the application space.

The blob storage can store file up to a size of 1 TB, which makes it particularly well suited for data-intensive application. The Amazon S3 service, e.g., restricts the maximum file size to 5 GB. Further, the access to the blob storage can be optimized for certain usage modes: *block blob* can be split into chunks that can be uploaded and downloaded separately and in parallel. Thus, block blobs are well suited for uploading and streaming large amounts of data. *Page blob* manage the storage as an array of pages. Each of these pages can be addressed individually, which makes page blobs a good tool for random read/write scenarios. *Azure XDrive* provides a durable NTFS volume, which is backed by a page blob. In particular, legacy applications that heavily utilize file-based storage can simply be ported to Azure using XDrive.

The Azure Queue Service provides a reliable storage for the delivery of messages within distributed applications. The queue service is ideal to orchestrate the various components of a distributed applications, e.g., by distributing work packages or collecting results, which could be running on Azure or on another resource, e.g., a science cloud.

The Azure Table Storage is ideally suited for storing structured data. Unlike traditional relational database systems, the table storage is designed with respect to scale-out, low cost, and high performance similar to Google's BigTable [14] system. For legacy application, Azure also provides an SQL-Server-based, relational datastore called SQL Azure. In contrast to Azure tables, SQL storage supports common relation database features, such as foreign keys, joins, and SQL as query language.

### 9.7.2 Azure: Understanding the Applications

The Azure platform provides many capabilities and characteristics that are useful for scientific applications. Clouds like Azure are particularly well suited for loosely coupled applications that demand a large number of processors but do not require a low-latency interconnect.

Although loosely coupled ensemble-based simulations are computationally well suited for cloud infrastructures, coordinating multiple ensemble members remains a challenge, as does data management. In the following we discuss a generic Azure-based architecture that addresses these concerns and is able to facilitate a range of applications execution scenarios.

**Azure and Bio-EnMD**

Figure 9.8 illustrates an example of a framework for ensemble-based simulations built on top of the Azure building blocks. The Replica Manager (RM), also called Ensemble Manager, utilizes a web role to communicate with the end-user. This role is mainly responsible for accepting simulation requests from the end-user and for orchestrating the simulation runs. Later these capabilities can be extended by supporting more advanced steering and visualization features. The RM creates work packages, the so called *replicas*, distributes them via the Azure Queue Service and later

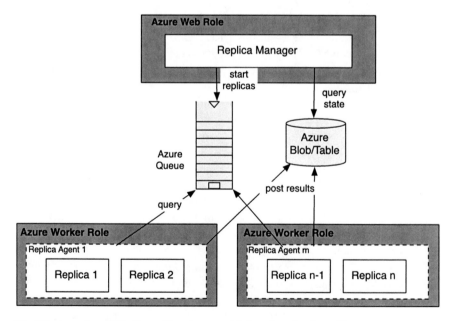

**Fig. 9.8** Azure-based Bio-Ensembles: the ensemble-based application utilizes the queue storage for distributing work packages from the Replica Manager running on a web role to the replica agents running on multiple worker roles

collects the results stored by the replicas in the Azure storage. The Replica Agents run within Azure worker roles, which are ideally suited for running background tasks. Azure enables users to run native code within worker roles, i.e., the framework will be able to support numerous MD codes, e.g., NAMD [31] and AMBER [13]. Azure currently does not support MPI computations across multiple worker roles, and thus each MD simulation is limited to eight cores.

The worker roles running the replica agent are managed by the RM using the Azure Service Management API. In the initial version we will support the automatic start and stop of hosted services. In the final version there will be a possibility to automatically deploy agent code without the need in preconfiguring the VMs. Once the agents are started, they query the Azure queue for new work items. If a work item is found, a simulation task is started, e.g., by running the requested MD code with right parameters. The number of MD jobs per worker role depends on the size of the worker role—Azure currently supports worker roles up to eight cores.

The ensemble use case can greatly benefit from the capabilities of Azure. If greater accuracy is required or a deadline must be met, it can seamlessly scale-out to more worker roles. The Azure fabric controller monitors all VMs running the worker roles and automatically restarts the worker roles if necessary. Further, Azure provides various kinds of reliable and scalable storage options to express different coordination schemes, e.g., the master–worker communication can be conducted via the described message queue.

**Azure and MapReduce**

For data-intensive applications, such as MapReduce -based applications, Azure provides various interesting services: The blob storage is well suited for storing large amounts of file data, which can also be accessed via an NTFS file system. Data can be processed via worker roles. Further, Azure offers the possibility to express data/compute colocations using so-called affinity groups. MapReduce application require capabilities similar to the Bio-EnMD: mapping and reduction tasks can be spawned dynamically using the Service Management API. For storing and transferring data, Azure provides various options (xDrive versus block blob versus page blob). We will further evaluate these alternatives in particular in comparison to distributed file systems in the future.

### 9.7.3 Assessing Azure-System: Abstractions and Applications

To aid an understanding of application characteristics suited to the Azure platform, we briefly introduce a common categorization of cloud services (see Fig. 9.9). The proposed service layers consist of the following: the software as a service layer (SaaS), the platform as a service (PaaS) layer, and the infrastructure as a service (IaaS) layer. Further, clouds can also be classified according to their deployment

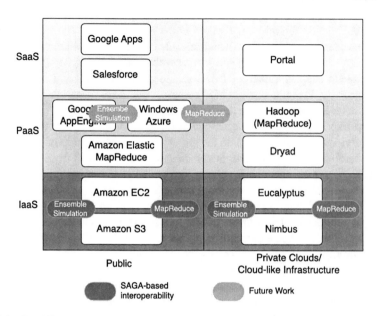

**Fig. 9.9** Cloud Taxonomy and Application Examples: Clouds provide services at different levels (IaaS, PaaS, SaaS). The amount of control available to users and developers decreases with the level of abstraction. According to their deployment model, clouds can be categorized into public and private clouds

model into public and private clouds (for further details, refer to [22]). Azure offers services on the platform as a service layer and thus, generally removes the need to manually manage low-level details as virtual machine configurations, operating system installations and updates, etc. At the same time worker roles provide an attractive environment for running compute- and data-intensive applications. In contrast to IaaS clouds, applications can benefit from features, such as failure tolerance: the fabric controller, e.g., monitors all applications running in a role environment and restarts them if necessary.

The majority of scientific applications (e.g., applications from life sciences, high-energy physics, astrophysics, computational chemistry) that have been ported to cloud environments (see [10, 19, 29] for examples), rely on low-level IaaS cloud services and solely utilize static execution modes: A scientist leases some virtual resources in order to deploy their testing services. One may select different number of instances to run their tests on. An instance of a VM is perceived as a node or a processing unit. In contrast to traditional IaaS clouds, Azure provides different benefits: Azure operates on a higher level of abstractions and removes the need to manage details, such as configuration and patching of the operating system. Azure applications are declaratively described and packaged; the fabric controller automatically handles the mapping of these applications to available hardware.

Azure provides several core services supporting various interesting application characteristics and patterns. Compute-intensive tasks can naturally be mapped to worker roles, while the communication and coordination between these roles is

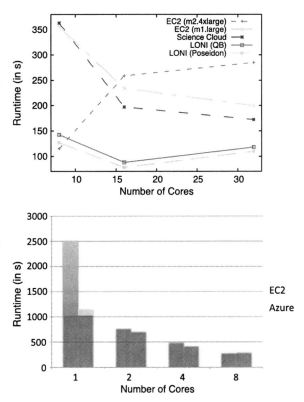

**Fig. 9.10** (a) NAMD Runtimes on Different Resource Types: The graph shows that native HPC resource generally outperform cloud resources in particular when running applications across multiple nodes. However, the new high-memory eight-core EC2 instance type was able to complete a replica run faster than QB or Poseidon. (b) NAMD Performance on Azure and EC2: In particular, on smaller VM sizes, Azure outperforms EC2. On eight-core VMs, EC2 shows a slightly better performance

commonly done via the Azure storage services. Worker roles can run not only .NET code, but are also capable of executing native code. However, Azure imposes some limitations in the ability to scale up and out. The largest supported VM has eight cores, 14 GB of memory, and 2 TB of disk space. Further, MPI applications can currently not be run on Azure. While other IaaS clouds can run MPI jobs, the performance usually degrades significantly when running jobs across multiple VMs (see Fig. 9.10a).

There exist a large number of physical problems that do not require high-end HPC hardware and interconnects and can easily scale out on Azure benefiting from the ability to acquire and release resources on-demand. An increasing number of applications directly target distributed infrastructures instead of high-end machines. For example, ensemble-based molecular dynamics approaches utilize multiple sets of simulations of shorter duration instead of a single longer simulation to support a more efficient phase-space sampling. Also, such simulations often require the ability to acquire additional resources if, e.g., a certain simulation event occurs that requires the spawning of an additional replica. Such an application can greatly benefit from Azure capability to dynamically allocate resources on demand. This capability is also useful for applications where the execution time and resource requirements cannot be determined exactly in advance, either due to changes in runtime requirements or interesting changes in application structure (e.g., different solver with different resource requirement or a different workflow path [24]).

Most research has solely attempted to manually customize legacy scientific applications in order to accommodate them into a cloud infrastructure. Benchmark tests on both cloud infrastructures (EC2, Azure) where a VM does not cross physical nodes and conventional computational clusters indicated no significant difference in the performance as measured by execution (wall-clock) time and number of processors used. Figure 9.10b presents an initial performance assessment of Azure for MD simulations. Azure outperforms EC2, which is noteworthy, since the costs for 2-, 4-, and 8-core VMs are drastically lower on Azure. Since the underlying hardware is not known, one can only speculate about the reason. Microsoft controls the hardware in its data center and optimizes its custom-built Azure Hypervisor with respect to this hardware [25], which could be a reason for better performance. In summary, Azure offers a good price/performance ratio, in particular, in comparison with EC2.

For data-intensive applications, Azure provides several interesting storage options: xDrive offers file system access to the Azure storage service, which is in particular relevant for applications that manage file-based data flows The blob storage is capable of storing large amounts of data, a page blob, e.g., can store files up to a size of 1 TB. In comparison, Amazon S3 is only capable of storing file with a size of up to 5 GB, the Google Storage for Developer file size limit is at 100 GB. The blob storage supports two different data access patterns: block blobs are designed for continuous access, such as data streaming, while page blobs can address each page individually and are particularly well suited for random access. These properties can be mapped to the characteristics of the respective application, e.g., a MapReduce application usually accesses data in large chunks, which is well supported by the block blob. In future we will investigate these implementation alternatives and performance trade-offs in conjunction with the proposed applications as part of this project.

## 9.8 Discussion and Conclusions

The aim of this chapter has been to show several types (and levels) of interoperability; although driven by proof-of-capability experiments and results therein, there are deeper questions that motivate this work and define the research methodology. As alluded to in the opening section, the volume and the degree-of-distribution of data is increasing rapidly; this imposes a need for applications to work across a range of distributed infrastructures using several programming models; this is consistent with the fact that it is not possible to localize exa-bytes of data. Thus, on the one hand, there is a need to decouple PM from infrastructures and provide a range of PM at the application developer's disposal. On the other hand, in order to build empirical models or validate existing predictions of performance, it is important to establish and experiment with programming models and data-oriented algorithms (e.g., streaming) on a range of systems. A critical and necessary step to achieve both is to provide application-level interoperability as discussed.

In this chapter, we discussed two important (classes) of applications, data-intensive SAGA–MapReduce wordcount and compute-intensive ensemble-based molecular-dynamics simulations. We posited three levels of interoperability and, for SAGA–MapReduce wordcount, carried out performance tests at all three levels. At the lowest level, SAGA–MapReduce demonstrates how to decouple the development of applications from the deployment details of the runtime environment (Type I ALI). It is critical to reiterate that using this approach, applications remain insulated from any underlying changes in the infrastructure—not just grids and different middleware layers, but also different systems with very different semantics and characteristics, whilst being exposed to the important distributed functionality.

With implementations of the two application frameworks, SAGA-based Sector–Sphere and the SAGA–MapReduce implementation, we also demonstrated Type II ALI: applications can seamlessly switch between backends by switching frameworks encapsulating different programming models. Finally, by concurrently using Sector–Sphere MapReduce and SAGA–MapReduce, we demonstrated Type III ALI, allowing the application to span a wide variety of backends concurrently and efficiently.

Our approach does not confine us to MapReduce and applications based upon MapReduce; SAGA is also capable of supporting additional programming models, like Dryad. We are also developing applications with nontrivial data access, transfer, and scheduling characteristics and requirements, and deploying them on different underlying infrastructure guided by heuristics to seek optimized performance. This analysis is done through developing performance models of transferring data between frameworks, as well as the distribution of the computing resources in the environment. Based on this analysis, the data is placed efficiently, and a subset of nodes and frameworks maybe chosen to perform the necessary computations. The shuffled data is also cached for future computations. We have embarked on the creation of components that facilitate intelligence and flexibility in data placement relative to the computational resource [9]. These components are connected in frameworks using SAGA, and thus further the agenda of general-purpose programming models with efficient run-time support that can utilize multiple heterogeneous resources.

In Sect. 9.6, we showed how SAGA can be used to develop frameworks such as infrastructure-independent Pilot-Jobs and demonstrated the scaling-out over multiple distinct resources. Finally, in Sect. 9.7, we discussed Azure, some of the system-level abstractions that it provides and analyzed how these can be utilized for ensemble-based molecular dynamics simulations. We anticipate significant activity in both scaling-up the use of Azure (for both applications discussed here and other novel applications) and integrating Azure with other infrastructure via the use of SAGA in the near future.

**Acknowledgements** SJ acknowledges UK EPSRC grant number GR/D0766171/1 for supporting SAGA and the e-Science Institute, Edinburgh for the research theme "Distributed Programming Abstractions." SJ also acknowledges financial support from NSF-Cybertools and NIH-INBRE Grants, while ME acknowledges support from the grant OTKA NK 72845. We also acknowledge internal resources of the Center for Computation & Technology (CCT) at LSU and computer resources provided by LONI/TeraGrid for QueenBee. We thank Chris Miceli, Michael Miceli, Katerina Stamou, Hartmut Kaiser, and Lukasz Lacinski for their collaborative efforts on early parts of this work. We thank Mario Antonioletti and Neil Chue Hong for supporting this work through GSoC-2009 (OMII-UK Mentor Organization).

# References

1. Jha, S., Merzky, A., Fox, G.: Clouds provide grids with higher levels of abstractions and support for explicit usage modes. Concurr. Comput. Pract. Eng. **21**(8), 1087–1108 (2009)
2. Jha, S., et al.: Design and implementation of network performance aware applications using SAGA and Cactus. In: IEEE Conference on e-Science 2007, Bangalore, pp. 143–150 (2007). ISBN:978-0-7695-3064-2
3. Jha, S., et al.: Developing adaptive scientific applications with hard to predict runtime resource requirements. In: Proceedings of TeraGrid 2008 Conference (Performance Challenge Award)
4. SAGA Web-Page. http://saga.cct.lsu.edu
5. Protocol Buffers. Google's Data Interchange Format. http://code.google.com/p/protobuf
6. NIMBUS. http://workspace.globus.org/
7. http://www.loni.org/
8. http://opencloudconsortium.org/testbed/
9. Miceli, C., Miceli, M., Rodgriguez-Milla, B., Jha, S.: Understanding performance implications of distributed data for data-intensive applications. Philos. Trans. R. Soc. Lond. Ser. A (2010)
10. Bégin, M.-E., Grids and clouds—evolution or revolution. https://edms.cern.ch/file/925013/3/EGEE-Grid-Cloud.pdf (2008)
11. Borthaku, D., The Hadoop distributed file system: architecture and design. Retrieved from http://hadoop.apache.org/common/ (2010)
12. Casanova, H., Obertelli, G., Berman, F., Wolski, R.: The AppLeS parameter sweep template: User-level middleware for the Grid. Sci. Program. **8**(3), 111–126 (2000)
13. Case, D.A. III, Cheatham, T.E., Darden, T.A., Gohlker, H., Luo, R. Jr., Merz, K.M., Onufriev, A.V., Simmerling, C., Wang, B., Woods, R.: The amber biomolecular simulation programs. J. Comput. Chem. **26**, 1668–1688 (2005)
14. Chang, F., Dean, J., Ghemawat, S., Hsieh, W.C., Wallach, D.A., Burrows, M., Chandra, T., Fikes, A., Gruber, R.E.: Bigtable: a distributed storage system for structured data. In: OSDI '06: Proceedings of the 7th USENIX Symposium on Operating Systems Design and Implementation, p. 15. USENIX Association, Berkeley (2006)
15. Cloudstore. Cloudstore distributed file system (formerly, Kosmos file system). http://kosmosfs.sourceforge.net/.
16. DeCandia, G., Hastorun, D., Jampani, M., Kakulapati, G., Lakshman, A., Pilchin, A., Swaminathan, S., Vosshall, P., Vogels, W.: Dynamo: Amazon's highly available key-value store. SIGOPS Oper. Syst. Rev. **41**(6), 205–220 (2007)
17. Deelman, E., Singh, G., Livny, M., Berriman, B., Good, J.: The cost of doing science on the cloud: the Montage example. In: SC '08: Proceedings of the 2008 ACM/IEEE Conference on Supercomputing, pp. 1–12. IEEE Press, New York (2008)
18. Nurmi, D., et al.: The Eucalyptus open-source cloud-computing system. October 2008
19. Evangelinos, C., Hill, C.N.: Cloud computing for parallel scientific HPC applications: feasibility of running coupled atmosphere–ocean climate models on Amazon's EC2. In: Cloud Computing and its Applications (CCA-08) (2008)
20. Ghemawat, S., Gobioff, H., Leung, S.T.: The Google file system. ACM SIGOPS Oper. Syst. Rev. **37**(5), 43 (2003)
21. Gu, Y., Grossman, R.L.: Sector and Sphere: the design and implementation of a high-performance data cloud. Philos. Trans. R. Soc. Lond. Ser. A **367**, 2429–2445 (2009)
22. Jha, S., Katz, D.S., Luckow, A., Merzky, A., Stamou, K.: Understanding scientific applications for cloud environments. In: Cloud Computing: Principles and Paradigms. Wiley, New York (2010)

23. Kaiser, H., Merzky, A., Hirmer, S., Allen, G.: The SAGA C++ reference implementation. In: Object-Oriented Programming, Systems, Languages and Applications (OOPSLA'06)—Library-Centric Software Design (LCSD'06), Portland, OR, USA, 22–26 October 2006
24. Kim, H., el Khamra, Y., Jha, S., Parashar, M.: Exploring application and infrastructure adaptations on hybrid grid–cloud infrastructure. In: First Workshop on Scientific Cloud Computing (Science Cloud 2010). ACM, New York (2010)
25. Krishnan, S.: Programming Windows Azure. O'Reilly Media, New York (2010)
26. Lu, W., Jackson, J., Barga, R.: AzureBlast: A case study of developing science applications on the cloud. In: First Workshop on Scientific Cloud Computing (Science Cloud 2010). ACM, New York (2010)
27. Luckow, A., Jha, S., Merzky, A., Schnor, B., Kim, J.: Reliable replica exchange molecular dynamics simulation in the Grid using SAGA CPR and Migol. In: Proceedings of UK e-Science 2008 All Hands Meeting, Edinburgh, UK (2008)
28. Luckow, A., Lacinski, L., Jha, S.: Saga BigJob: an extensible and interoperable pilot-job abstraction for distributed applications and systems. In: The 10th IEEE/ACM International Symposium on Cluster, Cloud and Grid Computing (2010)
29. Merzky, A., Stamou, K., Jha, S.: Application level interoperability between clouds and grids. In: Workshops at the Grid and Pervasive Computing Conference, GPC '09, May 2009, pp. 143–150 (2009)
30. Miceli, C., Miceli, M., Jha, S., Kaiser, H., Merzky, A.: Programming abstractions for data intensive computing on clouds and grids. In: 9th IEEE/ACM International Symposium on Cloud, Cluster Computing and the Grid, CCGRID'09, May 2009, pp. 478–483 (2009)
31. Phillips, J., Braun, R., Wang, W., Gumbart, J., Tajkhorshid, E., Villa, E., Chipot, C., Skeel, R., Kale, L., Schulten, K.: Scalable molecular dynamics with NAMD. J. Comput. Chem. **26**, 1781–1802 (2005)
32. Goodale, T., et al.: A simple API for grid applications (SAGA). http://www.ogf.org/documents/GFD.90.pdf

# Glossary

**Cloud Computing** a business model that provides utility Computing services and/or SaaS services.

**Grid Computing** distributed computing that enables IT scalability and flexibility, mainly focusing on large-scale problems.

**Platform as a service** Pay-per-Use for network-based delivery of a computing platform and a related solution stack as a service.

**Service Orientation** a design paradigm that specifies the creation of automation logic in the form of services. It is applied as a strategic goal in developing a service-oriented architecture (SOA).

**Software as a Service** Pay-per-Use for network-based software applications services.

**Utility Computing** Pay-per-Use for network-based Compute and Storage services.

**Virtualization** a technique for hiding the physical characteristics of computing resources from the way in which other systems, applications, or end users interact with those resources.

M. Cafaro, G. Aloisio (eds.), *Grids, Clouds and Virtualization,*
Computer Communications and Networks,
DOI 10.1007/978-0-85729-049-6, © Springer-Verlag London Limited 2011

# Index

**A**

Abstractions, 201
Accounting, 157
Advance reservation, 144
Agenda, 144
Amazon EBS, 81
Amazon EC2, 56, 63, 76, 179, 186, 191
Amazon S3, 76, 81
Application resource tuple, 25
Azure, 202

**B**

Billing, 157
Broadband, 77, 78, 83

**C**

Capping, 148, 151
Carbon footprint, 144
Client-server, 9
Cloud computing, 6, 169
Condor, 75, 76, 80
Condor glideins, 81
Consolidation, 145
Contextualization, 75, 76
Coordinated resource sharing, 5
Corral, 81
COST action IC0805, 171
CPU idling, 147
CPU throttling, 145
Cyberinfrastructure, 72, 76

**D**

DAGMan, 75, 80
Data-compute affinity problem, 15
DRMAA, 66
DVFS, 145
Dynamic voltage and frequency scaling, 145

**E**

EGEE, 58
EGI, 171
Elasticity, 73
Electricity cost, 144
Energy, 144
Energy awareness, 144, 155
Energy consumption, 144
Energy efficiency, 61, 144
Energy saving, 144
Energy-aware cloud framework, 144, 153
Ensemble-based simulations, 202
Enterprise application workloads, 24
Epigenome, 77, 78, 84
Everything as a service, 163

**F**

Fault-tolerance, 173, 176, 178, 180
Federation, 58, 64

**G**

Globus toolkit, 66, 81, 177, 191
GlusterFS, 76
GPFS, 76
GPU, 170, 174, 188
Green open cloud, 144, 153
Green policies, 157
Green SLA, 157
Grid computing, 5
GridWay, 60, 66

**H**

Hardware abstraction layer, 11
Hybrid cloud, 56, 63
Hypervisor, 147

M. Cafaro, G. Aloisio (eds.), *Grids, Clouds and Virtualization*,
Computer Communications and Networks,
DOI 10.1007/978-0-85729-049-6, © Springer-Verlag London Limited 2011

**I**

I/O device sharing bottlenecks on virtualized
    server, 35
I/O device virtualization challenges, 24
I/O virtualization architecture description, 38
I/O virtualization architecture wish-list, 38
IaaS, 56, 57, 61, 66
Ibis, 174
Ibis programming models, 175
IbisDeploy, 178
Illusion of infinite resources, 73
Infrastructure as a service, 72
Interoperability, 9, 173, 200
Interoperation, 64
IPL, 174

**J**

JavaGAT, 177
Join-elect-leave (JEL), 175
Jorus, 186
Jungle Computing, 171
Jungle Computing System, 169, 171

**L**

Legacy applications, 74
Legacy codes, 174, 181
Live migration, 150
Lustre, 76, 82

**M**

Makespan, 82
Malleability, 173, 176, 178
MapReduce, 193
MapReduce,Dean, 13
MAQ, 78
Metacomputing, 2
Middleware, 3
Middleware independence, 173
Migration, 144
Montage, 77, 83
Multimedia content analysis, 181

**N**

NCSA, 72, 76
Network presence, 155
Network QoS evaluation on virtualized server,
    33
Neuroinformatics, 183
NFS, 76
Nimbus context broker, 75, 88

**O**

OASIS, 9
OCCI, 63

On-demand, 73
Open grid forum, 9
Open science grid, 72
OpenNebula, 63, 65

**P**

Panasas, 76
Parallelization, 173
PBS, 76
Peer-to-peer, 9, 169, 180
Peer-to-peer middleware, 178
Pegasus, 75, 80
Pilot job, 59
Pinning, 148, 152
Platform as a service, 72, 89
Platform-as-service, 7
Power management, 154
Prediction algorithms, 157
Private cloud, 56, 60
Programming models, 201
Provenance, 74
Provisioning, 73
Proxy, 155
Public cloud, 56, 62, 64
PVFS, 76

**R**

Remote sensing, 184
Replica location service, 89
Reproducibility of scientific results, 74
RESERVOIR, 58
Resource independence, 173
Resource management system, 146
Resource manager, 154
Resource provisioning, 73, 145
Review of I/O virtualization techniques, 36
Robust connectivity, 173

**S**

SAGA, 201
Scale-out, 57, 63
Scaling, 3
Scheduler, 159
SDSC, 72
Semantic web, 182
Server consolidation, 60
Service level agreement, 9, 157
Service-oriented architecture, 72
SLA, 157
SmartSockets, 176
Software as a service, 72, 89
Software-as-service, 7
StratusLab, 58, 63

Sun grid engine, 76
System virtualization, 24

**T**
Temperature-aware scheduling, 145
TeraGrid, 72

**V**
Virtual clusters, 75
Virtual CPU, 147
Virtual machine, 10, 144
Virtual machine energy cost, 148
Virtual machine monitor, 11
Virtual organizations, 5

Virtualization, 72, 74, 145, 180
Virtualization and application performance, 28
VM, 144

**W**
Web application, 144
Workload consolidation, 145

**X**
Xaas, 163
Xen, 147

**Z**
Zorilla, 178

CPSIA information can be obtained at www.ICGtesting.com
Printed in the USA
LVOW082128141211

259490LV00005B/49/P